Your Tractor Purchase, Care...

Selected excerpts from this practical handbook:

Should I Rent or Buy?

"If you must rent, do a thorough inspection of the equipment before you leave the rental company and be sure to ..."

How to Inspect a Used Tractor

"I have heard hundreds of people tell me they wouldn't buy a used tractor with a paint job..."

Cooling System

"If somebody tries to sell you some specially formulated anti-freeze for your tractor for approximately $25.00 a gallon, tell them your Unicorn ate your wallet!"

Hard Starting

"Sometimes the solenoid switch can stick. You can dislodge it by ..."

Ford

"I'm surprised at how many people still don't realize that a New Holland is not a Ford ..."

Allis Chalmers

"The only way to fix a Long is to melt it down to a big blob and use it for a shrimp boat anchor..."

Gray Market Tractors

"I can't help you after you have made a mistake of this magnitude ..."

Your First Tractor

Your First Tractor: Purchase, Operation and Service

by
Gary Brown
Rudy Brown
Clay Brown

The First-Time Tractor Buyer's Self Defense Manual

YourFirstTractor.com

Published by
Starve Monkey Press, Inc.

StarveMonkeyPress.com

Copyright © 2009 by Gary Brown, Rudy Brown and Clay Brown
All rights reserved.

Unless otherwise stated, text and diagrams written and copyrighted by Starve Monkey Press or its authors remain the sole property of Starve Monkey Press or its authors, respectively. You may not modify this material or distribute it, in whole or in part, in any form whatsoever, to any other party, without written permission of Starve Monkey Press.

Published in the United States of America.

ISBN: 978-0-9825431-2-2
Library of Congress Control Number: 2010920826

Publisher Information

Starve Monkey Press, Inc.
Roberta, Georgia

See our website for more information about the publisher, including resources for aspiring authors and contact information:

www.StarveMonkeyPress.com

Trademark Notices

Starve Monkey Press and the SMP logo are the trademarks or registered trademarks of Starve Monkey Press, Inc. Shade Tree Conversions is a trademark or registered trademark of Gary Brown, *et al.* All other trademarks not specifically listed are trademarks or registered trademarks of their respective owners.

Version History

See the website for this book for errata and version history:

YourFirstTractor.com/errata

Original Paperback
First Printing, January 2010

The information expressed in this book is solely the opinions of the authors, and is for entertainment purposes only.

Table of Contents

Introduction 1
Should I Rent Or Buy? 2
Choosing The Right Tractor For Your Job 4
Selecting Between New Or Used Tractors 7
Should I Buy a Zero-Turn Mower Instead? ... 10
How to Inspect a Used Tractor 11
2- or 4-Wheel Drive? 17
Shuttle Transmissions 19
Standard Transmissions 22
Parking Brakes 25
Cooling System 26
Hard Starting 28
Safety Considerations 31
Maintenance and Service 41
Protect Your Tractor From Weather 48
Good Tips Before Starting 50
Tractor Brands 51
Ford 52
Massey Ferguson 67
John Deere 75
International - Farmall - Case 77
Kubota 81
Allis Chalmers 88
Yanmar 90
Long-Fiat 95
White Oliver 96
Belarus 97
Gray Market Tractors 98
Conclusion 101

Introduction

Buying a farm tractor, new or used, unless you are familiar with tractors, is a little like walking through a mine field blindfolded. Be careful where you step and turn. The big "gotchas!" are waiting for you. If you read this book, you will have the 100-plus years of knowledge and experience of the authors and you will be able to make an intelligent decision. This book could save you thousands of your hard earned dollars. This book is a must-read for people who have never owned a tractor before.

In today's market buying a used tractor is much more complicated than it was even 40 years ago. Almost everything in the equipment lineup back then was quality; American-made, and had a track record that spoke for itself. But today, especially, if you are considering a new tractor, the market is flooded with what we believe are substandard, low quality, and foreign-made tractors. Today's current model tractors are less durable products than tractors sold here before. They also have few or no available parts and supplies; and the resale value of a dead mule.

To make it even worse, you can't just trust the familiar brand names of the past, for instance John Deere, International, Ford (New Holland), Massy Ferguson, etc. Many of the big brand names of our past are selling something bought in Europe, China, India, South Korea or other countries, and are what we consider to be substandard at best.

So even if you have decided to buy a familiar brand name product, you can still be in trouble. This book will help you decide what the best value is for your hard earned dollar.

The authors were raised on farms and have been in the equipment business for over thirty years. What we learned the hard way, you can learn for a few bucks. On the other hand, don't let our opinions in this book, and that's all they are, our opinions, keep you from applying your own good judgement in any particular situation.

That said, tractors are good investments, if you buy the right one. They can also keep your property value up.

Should I Rent Or Buy?

If you own property in the mountains and only spend a week per year there, then you might be better off renting a tractor. When you rent, you take the responsibility of the particular tractor and equipment as soon as you leave the rental dealership. If anything breaks on the tractor or implements, you are responsible.

Today, more and more people are telling me horror stories about returning a rented tractor and thinking everything is fine only to get a big repair charged to their credit card. Remember you have to have a credit card to rent something. I talked with a rental dealer about the sudden increase in unexpected charges, and he defended the industry.

He said, "I don't think there was any corruption going on, but rental companies, because of the slower economy, weren't turning their inventory in for new equipment as regularly as they did before."

He then explained, "The older equipment with more hours and wear and tear tend to have more mechanical failures."

Regardless of whether the charges are corrupt or whether the equipment has been over worked, the fact still remains: if you rent, it's like musical chairs. If the music stops while you're renting something, it's "ouch" from your wallet. Many people don't realize that they are going to pay for repairs on a piece of equipment that has been rented by hundreds of individuals, many of whom don't know how to properly use the equipment and others who could care less. So, be careful.

If you must rent, do a thorough inspection of the equipment before you leave the rental company and be sure to have it posted on your rental contract if you see anything suspicious. If for any reason the rental company refuses to make special notes about possible existing damage or mechanical problems, take your credit card and leave.

It may be cheaper to hire your occasional work done. I suggest you get a price up front, and insist that if they break their tractor they agree to pay to fix it.

But, the average person will come out on top if they buy a good tractor and equipment, and they do the work themselves. In many cases you can write-off the price of your tractor on your income taxes, but consult a good certified public accountant or tax professional to be sure.

I built houses for ten years. I owned my own tractor for landscaping, and I hired backhoe work done. I paid enough money hiring the backhoe work done that I could have bought a new one and paid for it. If you buy a good quality tractor, that wouldn't necessarily mean a new one, it will more than pay for itself.

Often, people try to cheap out. They say, "I only need it a few times a year." If I've heard that once, I've heard it a thousand times. It makes no difference how many times a year you're going to use it, you still need a good one.

If you buy a new "EL CHEAPO" tractor that was junk the day it was built, you have thrown your money away. If you buy a cheap junk used tractor, you lose. Many people want to buy a good used tractor with a mower and box scraper for about $5000.00. You might get an older gasoline engine tractor without power steering and some used implements for that kind of price, but for a couple of thousand more, you get a diesel with power steering and new implements. This equipment could last the rest of your life, and you could will it to your kids. If they take care of it, they could leave it to your grandkids.

So if you rent or buy, be careful. If you have to read this book several times, it's OK. There is a lot of information in this book. I wrote this book to help people. I know some people don't want to be helped, and some can't be helped.

Choosing The Right Tractor For Your Job

If you have never owned a tractor before, it's OK to ask a friend or relative their opinion, but be careful. Only take advice from someone who has actually owned a tractor. Too many people take advice from somebody who had a cousin whose next door neighbor's friend knew somebody that might have owned one.

The first decision is what size tractor you need for your particular work. In the many years we have been in the tractor business, I have been amazed at what inexperienced tractor buyers think a tractor will do.

Many people believe that any size tractor will pull anything that you can chain to it. I honestly believe that some people think that if you could find a chain long enough to circle Stone Mountain, you could take a small 20 hp compact diesel tractor and pull it to Savannah. They believe any tractor will pull anything. They think little, small ones can do the same work as a big one. They also think that you could buy a little, tiny, single axle trailer to haul your tractor on, or you can just use the same trailer you use for hauling your lawnmower.

Little, small, compact diesels are only good for light work! Cutting grass with a finishing cutter or maybe pulling a small scrape blade is about all they will do. I guess you could pull a small trailer around your property with it and pick up limbs after a storm. These tractors, under approximately 30 hp, are not powerful or strong enough to have a loader or backhoe. In Japan where most of these are made (regardless of what name plate is on it) they don't put loaders or backhoe attachments on their small compact tractors.

As I opened up my shop one morning, an older Japanese man was standing out front on my lot. He told me that when he was young he helped design this particular tractor (a 750 John Deere) with a loader and a backhoe attachment. He explained that in Japan nobody would dare put these attachments on this tractor. He said they weren't built strong enough to hold up under such heavy use. He told me that in Japan they built excavators for digging and machines like our skid steers with tracts for loaders.

Choosing the Right Tractor For Your Job

He explained that, at that time in Japan, metal was a very precious and expensive material. He was warned if he used one piece of metal the size of a grain of sand that wasn't absolutely needed, even though he was an engineer, he would sweep floors the rest of his life.

So if you want to use a rough cut rotary mower, bottom plow, harrow, or post-hole digger, then you need a tractor around 40 horsepower or larger. The smaller compacts are generally priced less than the larger ones, and I think that's another reason some people tend to purchase the smaller tractors.

In my thirty plus years of selling tractors, I can count on one hand the number of people who traded a larger tractor back for a smaller one. Usually this would be a case where someone had fifty acres of land and sold all of it, except for one acre around their house. I can't count the number of people who came back and traded a small compact tractor for a larger one. They would lose approximately $1000-$2000 after scratching up their tractor and implements.

So, it is important to buy the proper size tractor for your needs. If you plan to move large round hay bales with the three point hitch or with a front end loader, I would suggest 50 horsepower or more.

I've seen many people who think they can put a loader on a small compact tractor and move their large round bales. They will eventually have a tragic accident or destroy the tractor, if they try this. At our small business we only install loaders on 40 horsepower or higher tractors and only on tractors with heavy duty front axles. Many of these tractors are the heavy duty or commercial versions of the standard agriculture tractor.

If you plan to operate a square baler, generally a 40 horsepower diesel will work well enough. But, if you plan to use a round baler, then you need to check with the manufacturer of the particular baler you plan to use. Generally speaking, a tractor of 70-80 horsepower will be better.

I need to warn you that if horsepower is your only guide, you need to be careful. Many of the Japanese made brands don't have the weight to horsepower ratio that you need. I like a ratio of 80-100 pounds of weight per horsepower. For instance, if you try to use a heavy 6 foot rotary mower on a 40 horsepower Kubota, you're in trouble. A 40 horsepower Ford or John Deere would weigh approximately 1000 pounds more. You need the proper weight for the horsepower.

We sold a 5600 Ford (65 horsepower) to one of our customers and after his ship came in financially, he traded his Ford for a new Kubota tractor of the same horsepower. Then he learned the hard way that he needed the extra weight. Even though the new tractor had the same 65 horsepower rating, the tractor couldn't pull any of his plows, harrows, or mowers. He had to trade the 65 horsepower tractor back to the dealer for a 90 horsepower Kubota to do the job.

So, first it is important to buy the right size tractor to suit the type of implements you need and make sure the tractor can hold up under such stress. Second, if you have a small single axle trailer and you are trying to buy a farm tractor to do heavy work, it's best to buy the proper tractor even if you have to trade your small trailer for a larger one.

Selecting Between New Or Used Tractors

After you have decided what size tractor you need and the necessary features you want, you're ready to go shopping. It's best to check out the new dealers first to get an idea of what the prices are for new tractors even if you plan to buy used. This will give you a perspective on the kind of money you'll spend for a new one. Then you need to investigate the price of used tractors. Most individuals who move out of town and buy a few acres in the country don't really need a new tractor. In most cases you're only going to use it approximately 50 hours a year or less. A farm tractor will complete a job in minutes compared to a riding lawn mower which would take hours or days to complete.

I will give you an important suggestion. Never, ever, walk through the door of a new or used tractor dealership and tell them you don't know anything about tractors, unless you know the dealer personally and trust him completely. Unfortunately there are people in this business who will take advantage of you if they know that you don't know anything about tractors.

The tractor market we have today is flooded with some of the worst junk on earth. When some dealers hear those words, they may try to stick you with a new or used basket case.

In my opinion you should read this book and talk to people who have owned a tractor for years. Never take advice from someone who doesn't own a tractor. In our opinion the world's sorriest tractors come from China. The saddest tractor buyers I've ever met just bought a Chinese tractor. You have to be very careful, because they are sold under so many names. I would run out of ink before I could list them all.

People who sell these tractors aren't usually in business very long and parts availability is almost nonexistent. In my opinion, unless the quality of these tractors improves tremendously, don't buy one. Many people have called us after owning one only a few days and told us about major mechanical failures. They want to know if we can fix them.

My response is, "If it only has a few hours on it and has already experienced a major mechanical failure, even if we could get the necessary parts to fix it back as good as new, it could have the same failure as before in just a few more hours."

If you buy a new or used tractor, regardless of the name on the hood, you want the seller to list on your Bill of Sale who made the tractor and where it was made.

Be careful when buying a used tractor from some "Bubba" who lives down the road. It's best to buy from somebody who has a good reputation and has been in business for many years. You have to worry when you buy from an individual about the unit being stolen or if the unit has liens on it. If you buy from an individual, you need to have the seller write on your Bill of Sale that the equipment is owned by them and free of liens or encumbrances. This means, if the tractor has a lien on it, that the seller (not you), will be responsible for the debt.

You have to be careful, because some people will sell a tractor for which they owe ABC Bank $5000. Then they sell it to you. In most states the bank could come after you. Simply write that phrase above, and don't forget the seller needs to sign the Bill of Sale in front of you. If they refuse, keep looking for a tractor. It's important when dealing with an individual to be sure the seller writes the name of the tractor on the Bill of Sale. He should be able to show you the serial number plate. This is like the VIN on a car. You should never accept a Bill of Sale that just says "diesel tractor." It must state what kind of tractor it is.

I had a man show up at my business one day to show me what a great deal he had found. He had an old French made 130 Massey Ferguson and a mower behind his pickup. He had told me the day before he could buy a Massey Ferguson 135 diesel and mower for $4250.00. The tractor on his trailer was a French tractor with no parts available. Even though it had a fresh coat of paint on it, it was in terrible condition. I wouldn't give $400.00 for it, much less $4250.00.

I explained to him that he had been "had." I asked him to look at his Bill of Sale. The man just looked blank.

He said, "The man changed the subject every time I asked about a receipt."

I asked, "Where did the man who sold you the tractor live?"

Again he looked blank. Finally speaking, he muttered, "We met at a shopping center across the state line in a small town."

I felt sorry for this man. It occurred to me this wasn't my fault. If the man had bought a tractor from us, he would not have gotten the shaft. Then I realized it's every businessman's job to protect the public. Most people in business would have just walked away and said, "Well, that's tough."

I suddenly remembered that I had just sold a tractor to a deputy sheriff from that very town. I raced inside my office to find the sales ticket to retrieve the deputy's name and phone number. In just a few minutes I had him on the phone. I described in detail what had happened. The deputy wanted to know the seller's name, but the victim told me all he knew was that his name was Buddy.

The deputy wanted a description of the seller and a description of his vehicle. "Bingo!" The officer recognized from the man's description who the seller was. He told us the individual was currently on parole for selling stolen equipment, for cheating, and swindling. The officer wanted to know if it was a cash transaction.

The buyer said, "Yes."

This man was lucky. The deputy sheriff got the buyer's money back by threatening to arrest the seller for "theft by conversion." But don't count on being that lucky. If this happens to you, your money will simply be gone.

Should I Buy a Zero-Turn Mower Instead?

Zero turn mowers are fine for cutting grass, but if you want to pull plows, cut rough stuff, drag a box blade or scrape blade, or anything other than cut grass, you're in trouble. In 95% of the cases you could buy a good used farm tractor with a rear mount finishing or grooming mower for about the same money.

If you buy a good tractor and finishing cutter, they will be running 40 years from now. The up-keep on a tractor and finishing cutter would generally be much less. If you buy the right brand tractor and service it properly and take good care of it, the value will actually increase.

My experience with zero turn mowers has shown me that the warranties run out much too soon. The tremendous number of moving parts is expensive to keep up, and the resale value drops like a brick.

I have also heard complaints from customers, especially people like me over 40 years old, suffering from neck and back pain due to the sudden turns. These machines also create large amounts of dust that can irritate allergies. Generally when you cut grass with a tractor and finishing mower, you leave the dust behind you as you move along. Lastly, the high pitch noise can damage your hearing.

So, for most mowing needs, I would suggest buying a tractor and finishing mower. If Aunt Sadie accidentally backed into a ditch, your zero turn mower would be useless.

How to Inspect a Used Tractor

Usually, the number one thing people are concerned with is the paint and the tires. If a tractor is painted, they think it's in excellent condition. They think that if a tractor has faded paint, it's junk. If the tires are old and dry cracked, they think the tractor is worthless.

The next thing they want to know is what year it was made. Only one in fifty people pull the engine oil dip stick and check the oil and then start the engine. This is the best step. The oil can tell you plenty. If it's low, it probably burns oil. This is a sign of engine wear.

If the engine has to turn over and over before it starts, that is a sign of low compression.

If the oil looks milky, it could be a busted engine block. It could be a blown head gasket, but few dealers would dare try to sell one like that. It doesn't cost that much to replace a head gasket.

A busted engine block on the other hand could cost several thousand dollars. If the oil looks multi-colored, that's a sign of oil additives that supposedly make a worn out engine run better. You can smell the oil. A strange smell means trouble. I like to wipe the oil from the stick with my fingers to check for all of the above problems.

The radiator cap could also spell trouble. A low coolant level could mean a coolant leak. Look closely at the coolant. Look for block sealer, which appears as a slimy silver or copper substance on the inside of the radiator neck, or on the bottom of the cap.

After starting the tractor, drive it around just like you would a used car. After a few minutes, practice starting, stopping, and making some right hand turns and left hand turns. This will test the clutch, steering, and brakes.

Stop the tractor. Take the tractor out of gear. Then, step off the tractor and walk around it. Look for leaks. Listen for rattles. Find the crank case vent pipe, which is a pipe or tube running from the valve cover down the side of the engine, and open on the end. Then, look for any smoke puffing out of it. This is a definite sign of engine wear. If the compression rings on the pistons are worn, when the cylinder fires some of the pressure escapes by the rings and goes out the crank case vent pipe, instead of exiting the engine through the open exhaust valve and then through the exhaust pipe and muffler.

Get down on your knees and look under the tractor. You need to look for leaks from the radiator, engine, transmission, and rear end. You should look for cracks or welds on the under side of the running gear.

All of these items mean trouble, so it's time to forget this one and move on.

But if the tractor has passed so far and while the engine is still idling, walk around to the rear of the tractor. Turn on the PTO to see if it starts and stops like it should. Then let the three point hitch down. I then like to stand on the lift arms and reach over the back of the tractor and lift the hydraulic lift lever to see if the lever lifts properly. I like to lift it just a little at a time to be sure it's working properly.

On our lot we generally have a mower, box blade, or some other implement on the back, so you can see for yourself how the lift functions.

Now that the engine has run long enough to warm up check out the exhaust. Look for smoke. Remember in damp, cool weather the exhaust will be more noticeable, but on a clear day, it should be clear.

Special note: If you suddenly rev-up a diesel, you will get a sudden puff of brown to gray smoke. This is normal, don't panic.

Now you should check the front end for excessive wear. If you turn the steering wheel back and forth, you can see how much wear is in the steering system.

You should also take into consideration the price and age of the tractor. If you are checking out an old 8N Ford with a price of $2000.00, some play in the steering and front end would be alright. If you're looking at a late model used tractor with a price tag approaching $10,000, you would expect it to have very little wear.

Now walk back to the rear of the tractor and examine the lift arms and the lift arm ball sockets. This is the part you hook your plows up to on the tractor. Test the adjusting control arm that allows you to raise or lower the lift arm on one side. Be sure it functions.

A more expensive used tractor, say one over $5000.00, should have headlights and a rear light that works. If your price tag is a $1500-$2500 clunker, consider it icing on the cake if *any* of them work! These are easily repaired at little cost.

Then, look for physical damage to the tractor such as warped or bent front axle steering rods and front axle stabilizer bars. Look for bent running boards and damage to the hood and sheet metal.

Continue to bear in mind, if you are looking at a $1500-$2500, 1950 or 1960 tractor, you can't expect it to be as straight, clean, and free of wear and tear as a more expensive one.

Some people have champagne taste and beer finances. The more you pay, the better the tractor should look and perform. Now get back on the tractor and drive it again modestly. Rev the engine a couple of times and watch the exhaust as before. Next check the gauges: oil pressure, and engine temperatures are most important. Then check the charging light, the tachometer, and hour gauge.

Again an older bargain basement unit will probably not have all the dash features working properly, but insist on temperature and oil pressure.

Now shut the tractor down. Walk around the tractor to the front to listen for radiator leaks and check under the bottom again. Warmer lubes and fluids will be more likely to leak.

Get back on the tractor and re-start the engine just to be sure it will start quickly and smoothly.

Let's discuss tires a little more in depth. Old, dry cracked tires on an older, lower priced tractor would be alright. My cousin, Tex Watson, is a house mover, and he explained to me why the "moving dollies" he uses has old dry cracked tires.

He said, "The old tires are much tougher, and they will roll over broken glass and junk and keep on going. The old rubber is tougher than new rubber. And those cracks are usually only surface blemishes.

Don't get me wrong. New tires are a bonus on a used tractor. I would be more concerned about tires with boots (another word for large patches), than I would dry cracks.

Now let us get back to the paint also. In our thirty plus years, I have heard hundreds of people tell me they wouldn't buy a used tractor with a paint job. Supposedly the customer is afraid a paint job will hide the tractor's faults or problems, but I've only sold a few people an unpainted tractor. I have found the lion's share won't buy a tractor that's not painted. I have had a much harder job trying to convince the average buyer that an unpainted tractor can be good.

I once got a call from a man that wanted a really good used tractor with a loader. I had just traded for a 340-A Ford industrial tractor with a factory Ford loader. It ran like a new one. It was strong, tight, had low hours, quality condition, but it also was very faded and covered in dust and dirt.

He arrived at our lot and explained what he had called about. I led him out back to where I had just unloaded the last load of new dealer trade-ins. The man was furious.

"I thought you said you had a nice one." He growled.

I tried to explain that he asked me if I had a real good one, not a freshly painted one. He got mad and left.

About a week later after we had washed, painted, and serviced this same tractor-loader, the same man walked through our gate and wanted the price of the same tractor-loader that he had turned his nose up at before.

I told him, "The price was $8750.00."

He said, "I'll take it."

Unfortunately I had to explain to him that it was already sold.

Then he yelled, "Why didn't you call me? You knew I wanted a good one like this!"

I replied, "This is the same tractor you saw just over a week ago. We just painted it, but you wouldn't listen when I tried to explain to you that this tractor was in excellent condition."

The man refused to believe it was the same tractor. About a week later I saw the same man hauling one of those old

Yanmars with a loader behind his pickup. I felt sorry for him, but I tried to tell him, but he wouldn't listen.

A month or so later I went by a friend of mine's house. He's retired, but works on tractors part-time at his workshop behind his house. I saw a Yanmar with a loader parked behind his shop. He told me it belonged to the same man who I had tried to sell the 340-A to. A rod had snapped and was protruding through the side of the engine block.

"I've been trying to find parts for this thing, but nobody seems to have them," he remarked.

The man had turned his nose up at an excellent 340-A Ford, because it wasn't painted. He then bought a worn out, sorry tractor, because it was painted; now his money was wasted. This was the final straw. WE decided to write a book and try to explain in plain English the confused jungle the tractor and equipment business has become. The more inexperienced you are, the more you need to read this book. And finally two of the most frequently asked questions are "How old is it?" and "What model year?"

Let me explain something. If I had a choice between a 1965 Ford 3000 in excellent condition or a 1995 Ford 3930 that was worn out and kicked to crap, for the same money, I would have no problem. I would take the 1965 Ford 3000. The year of a used tractor, especially today when so many new ones are junk, is much less important than the condition.

Banks call us wanting to know the year and value of a particular used tractor. We can usually find the year model in one of our books, but I have never found a book that could give me the true value of a used tractor or other equipment. I explain to them the value could only be made upon a person's inspection of the tractor or equipment. Most banks are too cheap to pay for an appraisal and just want to look in a book. SORRY, it just doesn't work like that. No two used tractors would have the same value. It's according to how these tractors have been operated, serviced, and how much wear is on them. It would also make a difference as to what they were used for.

One of the smartest men and one of the best men I've ever known was the late H. Lee Miller, a banker from Warner Robins, Ga. He taught me the true value of something is what it can be sold for! If you think about it, he had it figured out. I have a habit of quoting him, so I call him the "Chief." When it came to banking, he was the best. He proved to me that a man could possess honesty, integrity, wisdom, and character, yet still make lots of money.

As I wrap up this section, when dealing with a business always feel free to call the Better Business Bureau and talk to former customers and ask them how they were treated. Remember any warranties, guarantees, or promises of customer service are only as good as the people you are dealing with.

For a convenient checklist of the material in this chapter to take with you as you look at used tractors, go to our website and print our inspection form.

YourFirstTractor.com/inspection

You can share that form with your friends, too.

2- or 4-Wheel Drive?

We built the World Trade Center, Hoover Dam, and most of the world's largest projects without 4 wheel drive machinery. Yet today people, especially those buying smaller tractors, seem to think 4 wheel drive is mandatory. Some people think a 4 wheel drive, 25 horsepower machine will pull as much as a 2 wheel drive 50 horsepower machine. That is absolute nonsense!

Now don't get me wrong. Four wheel drive machines, when used properly, can be very helpful. For instance, 4 wheel drive is required under muddy conditions in a feed lot operation in the winter or in Alaska after the snow melts. But, in most parts of America, I don't believe 4 wheel drive is mandatory. When you have 4 wheel drive, you have a lot of extra moving parts to keep up. If you are buying a used tractor or tractor loader, be extra careful. My experience with 4 wheel drive has been 90% bad.

Unfortunately most new tractors with 4 wheel drive, because of the extra price, are purchased by rental companies, large corporations, or the government. My experiences with equipment from these sources have been bad. When people operate equipment that belongs to these particular groups, no one seems to care about how it is serviced. In many cases in the rental equipment business you have operators who don't know how to properly operate it, or who couldn't care less about how badly they abuse it. I would suggest that if you want a 4 wheel drive that you buy it new and maybe consider the extended warranty.

The worst problem I've seen is that these operators not only abuse this equipment, but these machines should only have the 4 wheel drive engaged when absolutely necessary. My experience from visiting thousands of job sites is that 98% of these units that I've checked have the 4 wheel drive engaged! Generally speaking, the 4 wheel drive should be engaged only on rare occasions when it is necessary.

In my opinion the 4 wheel drive part of your tractor running gear was not designed for continuous use. If you buy one used, you need to pay close attention to the loose play in the front end. I would suggest you ask the dealer to raise the front off the ground with a jack so you can check for loose play in the hubs, axles, and front drive shaft. Any loose play or noises could cost you a small fortune.

Most people believe that if the front wheel drive goes bad, you can just remove the drive shaft between the front drive unit and the transmission output shaft, but in most cases the problem will be either in the transmission or the front drive unit will lock or the front wheels themselves will lock up rendering your tractor useless. I suggest when buying used machinery skip the 4 wheel drive tractors unless it's absolutely mandatory.

Another fallacy I have found is that people think you have to have 4 wheel drive with a loader. It does help especially on the little compact tractors, but it's not mandatory. If you buy a 40 horsepower tractor, for instance a Ford, International, or John Deere (other than the new compact tractors), you have enough weight and horsepower to handle your job.

So my advice is, if you think you absolutely have to have a 4 wheel drive, buy a new one, or be prepared to cover the expenses of keeping a 4 wheel drive running.

Shuttle Transmissions

I'm not a fan of shuttle transmissions on a used tractor, compact tractor, or especially a compact tractor with a loader. On the smaller compact tractors the shuttles generally aren't much stronger than a hydrostat on a riding lawnmower. If you only intend to use your small compact tractor as a lawnmower it's OK, but be sure to completely stop the tractor with the brakes when shifting from forward to reverse.

I would never buy a new or used compact diesel tractor with a loader and shuttle transmission. I have a fear that new equipment dealers must be banking their retirement on the profits from the repairs of these units. I have been to large equipment auctions that often liquidate rental stock and checked out as many as 40 small compact tractor loaders with 4 wheel drive and shuttle transmissions. After close inspection I found all 40 to have at least one major problem with these options. Usually more than one.

I think because of the inexperience of the rental fleet operators and sometimes the intentional abuse at their hands, I would never buy a used unit like this. In some cases you will find a compact diesel tractor with a loader and backhoe with a shuttle transmission. I would run from one of these!

I would suggest, especially, on the compact tractor, if you are determined to buy a used one with a loader, buy a tractor without a loader and then have the dealer install a new loader for only about $1000.00 more. Then you'll have a tractor with much less wear and tear on the front end, steering, clutch, brakes, and transmission, plus a brand new loader for only about $1000.00 difference. For example, a used compact tractor without a loader will generally sell for approximately $2000-$3000 less, while a new loader installed will generally cost approximately $3500.00 installed.

Something I have noticed over the years is that most people who buy a tractor with a loader (unless you have a mandatory need for a loader) remove it within a year. Everybody seems to think of a hundred uses for one, but after a few months everything has been handled. Afterward, that loader, whether up or down, is just in your way. For that reason if you buy a used loader you can save a bundle.

I've often suggested that people buy a tractor and then run a wanted ad in one of those local Buy-Sell-Trade publications. Many of them run "wanted ads" for free. (P.S. with a wanted ad you are the only potential buyer. It's a buyer's market.) Then buy a good used loader, a couple cans of spray paint, and a little masking tape and you can make a good used one look good as new and save a lot of money.

Loaders can be very handy, but you need to be extra careful when you own or operate a tractor with a loader. When you load your bucket and drive your tractor, let the bucket down just above the ground until you get to where you are going. This lowers the center of gravity and helps to prevent an overturn. Also when the bucket is up in the air, you should pay close attention to the electrical service drop lines leading to your home or other out buildings. Another good point is to always leave the bucket on the ground when you park it. Kids are more fascinated by a tractor than the space shuttle. So always leave the buckets on the ground when parked or unattended.

You need to be very careful about the materials in the bucket. If you raise the bucket too fast and then turn the lever to tilt the bucket at the top, the items in the bucket could fall backwards toward you. When using those easy hook-up forklift attachments, use extra caution. If you buy a tractor that has any type of forks added to the bucket that aren't bolted securely, I suggest you throw them away.

Recently my brother was injured by a set of those forks on a 3400 Ford tractor loader. The loaders on the 3400 are really quick. When he raised it up and then rolled the bucket back to service the tractor, the steel bar with the two heavy steel forks on it unhooked and fell backwards. They then slid down the loader, pinning him by his legs. It's a miracle he only received a badly bruised leg muscle. It could have been much worst. We thanked The Almighty for sparing his life. We also have heard of similar accidents with those poorly designed forks, so I suggest you either bolt them on secularly or remove them completely.

If you operate a loader tractor moving large round bales of

hay, they can roll down the loader on top of you, so you need a good back-catch behind the forks to stop the bale from rolling over the back of the loader. For this reason I suggest removing the bucket and using properly mounted forklift attachments for these type jobs.

If your tractor has a (4) post top instead of a (2) post roll bar, you are somewhat safer, but you could still be injured.

My advice is, if you don't need a loader for a particular reason, then don't buy a tractor with a loader. If your tractor has a loader, be sure to keep a close eye on the mounting bolts, especially on the quick disconnect models. If a bolt falls out, you could damage your loader and/or tractor and you could be injured in the process. Safety is always the best policy.

My father used to say, "It only takes a few minutes to be safe, but an injury could last a lifetime." He was right, as usual.

I also suggest that you never install a loader on a tractor without power steering and a positive traction rear end. You need both of them to be able to steer the tractor when the bucket is full, and the positive traction will help you load the bucket on occasion. Again I can't overstate the fact that you need a sturdy front end.

I can't count the number of people who told me they didn't plan to put anything heavy in the loaders, then they loaned it to a neighbor who then pulled a 2 ton engine out of a semi truck. My dad always said, "Don't loan your tractor to anybody that you wouldn't loan your wife to."

About 90% of the repairs in our shop for physical damage happened when the tractors were loaned out to some friend or neighbor. I have found that people who borrow a tractor, in most cases, couldn't care less about the tractor, because it's not theirs.

My advice is that if you have a constant need for a tractor with a loader for farm or commercial purposes, buy the industrial version of a Ford like a 340 or 340A, etc.

Standard Transmissions

I strongly suggest a manual transmission, if you are going to do a lot of hard work, especially on the smaller tractors. The shuttle shift may be easier to operate, but when it fails, your wallet will feel much thinner. Some tractor dealers charge $4000-$8000 to repair them. A manual transmission would only need a clutch over time and could cost as little as $100 in parts or $200-$250 labor to replace.

I own a 1951 8-N Ford tractor with a 4 speed transmission and standard clutch. This tractor was bought used in 1957 by my father from McRae Ford Tractor Company. We were told it was bought new by the Talmadge Plantation, former home of our late US Senator. When dad bought it, he had them rebuild the engine, paint it, and completely service it. To the best of my knowledge, it has the original clutch and still works fine. This is a simple tractor and very reliable. I believe that a simple machine will outlast a dozen complicated or sophisticated ones.

Some of the newer high tech tractors have computers. A computer on a tractor is about as smart as a screen door in a submarine. Computers don't hold up well around heat, vibration, moisture and dust, so I never would buy a tractor with a computer. I have also become painfully aware that the new computerized line fire diesel engines burn a lot more fuel. I thought the reason for using diesels was to save fuel. I wonder if this isn't some sort of plot by the oil companies to destroy the fuel efficiency of our farm tractors. I have talked to farmers that have identical tractors, ones with the new computer model and ones with the standard models. The standard models generally run 2½ to 3 times as long on a tank of fuel.

We started a company, Shade Tree Conversions, in 2008 and installed farm tractor engines in cars and pickups. The Mustangs get up to 50 miles per gallon and the full size pickups get up to 40 miles per gallon. We sell books from our website at www.shadetreeconversions.com explaining how to install the tractor engines. Yet, some people, against our advice, used computerized engines. They learned the hard way to leave them alone. Ronald Reagan often repeated the old saying, "If it ain't broke, don't fix it." Wise advice.

The old-style diesels are the best on fuel efficiency, and you don't have to worry about buying a computer. Some cost upwards of $2000 every time you turn around. I like diesel engines, because they burn much less fuel, and they can run on bio-diesel, which is diesel mixed with used vegetable oil or fresh vegetable oil. You can mix diesel with used transmission fluid and used motor oil. Just filter it until it runs out good and clean.

When we published our Shade Tree Conversion book, the phone rang off the hook. People all over America were calling us to tell us that they were burning used transmission fluid and/or motor oil in their trucks and tractors. Many large trucking companies just filter the oil right out of the crank case and pump it into the fuel tank. They say this actually makes the trucks run better, because it replaces the lubrication that was lost when they removed sulfur from our diesel fuel.

Automobiles in Europe have diesel engines. Ford, Chevrolet, and Chrysler build identical vehicles there that get approximately double to triple the mileage that our American cars get. Ford has a 1.6 liter, 4 cylinder, diesel Focus built in Europe that gets over 60 miles per gallon. I believe the oil company money buys the press with their mighty advertising dollars, and they buy our congressmen and senators with the PAC money!

This nation will never survive unless we rise up and demand the PAC money paid to our politicians be shut off. I believe that every dollar they accept from these giant corporations will over time cost the American people hundreds of millions of dollars! I am worried about our country's economy. I believe that our nation was brought to its knees by the oil companies. From August of 2005 to September 2008, they robbed us of approximately 2 trillion dollars, and I believe that is why we are in the financial nightmare we are in now!

The news networks and politicians say it was the home finance business that caused this disaster. I disagree. The housing industry, the clothing industry, the furniture industry, and everything else fell when the oil companies robbed us blind. I believe the press and our politicians try to throw the blame on the housing industry to keep the focus off of them.

I love diesels. But, if you can find a good used farm tractor with a gasoline engine at a reasonable price, in many cases the older gasoline tractors won't have as much wear and tear on them. If you don't plan to run them every day, then I would say it was OK to buy it. When the 2000 and 3000 Ford tractors built between 1965 and 1975 were new, the diesel engine model was approximately $1000 more. This would be the equivalent, in today's market, if the gasoline model was $15,000, then the diesel model would be $20,000. If you planned to run one night and day, you could probably save the difference by buying the diesel. However, if you only planned to run it occasionally mowing the field behind the house twice a year and maybe working a garden, you would never save the difference. So in many cases the gas ones have much less wear and tear. The gasoline models will also usually sell for approximately $500-$1000 less. If you don't plan to use it a lot, I think buying the gasoline version would be OK.

If you plan to buy a used tractor older than 1965, I would prefer the gasoline engine over the diesel. These older diesels are very expensive to rebuild and they usually have double to triple the hours for the reason I explained before. Some of these older diesels have been converted from gasoline engines, but very few of them are still running.

Parking Brakes

It's a good idea to engage your parking brake when you leave your tractor for an extended length of time. I like to let the implements down on the ground and shut off the engine.

If the brakes on your tractor are hot, park the tractor and let the brakes cool. Then, reset the brakes. When the brakes cool, they contract or shrink and can cause your tractor to completely lose its grip, so you have to let them cool first before you can look the other way.

This can happen even after running it for only a few minutes if you have been using the brakes heavily, and it only takes another few minutes for the brakes to cool and release.

The tractor could roll so I always tell people "Don't trust the parking brake." To be safer, put the implements on the ground. When parked, also put the transmission into low range and low gear. That transmission and engine compression is a better brake than the parking brake.

I'm working on a book of country stories. There are more tractor brake stories in there than I would like.

Cooling System

The number one over-heating problem people bring to us is a dirty radiator, no matter how many times I tell people to be sure to keep the radiator clean. They always call us and want to bring it in for repairs.

Every time I ask them, "Are you sure the radiator isn't clogged with straw, leaves, bugs, dust, and debris?" Yet, 99% of the time when they get to our shop, we remove the grill and "Bingo". There is so much junk you can't even see the radiator. We simply take an air hose and blow out the trash and send them home.

You should take the grill out and clean the face of the radiator every time you spend the day mowing. You can wash it out with a water hose, if you don't have an air compressor. I don't like to spray cold water on a hot radiator. Many radiator repair shops will tell you it is OK, but they sell radiators. Just take the grill out or set it on the seat to remind you to do it the next morning.

The reason the debris sticks to the front of your radiator is because the engine fan on the front of your engine and right behind the radiator pulls air from the front of your tractor through the radiator to cool it. That is why it sticks to the front.

Never open the radiator cap when the engine is hot, especially if it is over-heating. You could get some really bad burns from the steam. Some people believe if you open the radiator when it's hot or overheating, it will cool down faster. The best way to cool your tractor would be to let the engine idle, provided the radiator is clean and has coolant. It should cool down in minutes.

If you check your coolant level regularly before starting the engine, and keep the radiator hose clamp tight, and, of course, make sure the front of the radiator is clean, you shouldn't have any over-heating problems.

If you begin to notice that the coolant is low, you need to examine the water pump. On most models the radiator cooling fan bolts to the water pump. If you see coolant dripping from the center of the fan, it's probably because the water pump is bad. Another way to check it would be to be sure the engine is off and just grab one of the fan blades. If it wiggles it's a good indication that the water pump bearing has gone bad and the seal will rupture right along with it. You should also check again when the engine isn't running to see if you can turn the fan blades by hand. This would tell you the fan belt is too loose.

Your tractor fan belts circle the crank shaft pulley, and then go around the water pump and around the generator or alternator; that is where you will find the adjustment. Just loosen the bolts on the generator or alternator bracket and pull out till the belt feels tight and tighten it.

It is normal for you to be able to flex the belt up to ½ inch, if you grab it between the pulleys. You need some slack in the belt. If you still have problems, it's probably the thermostat. It is generally located where the top radiator hose bolts to the head on the top front of your engine.

We give our customers service books with the tractors we sell, and most of our tractors are used. If you buy a new or used tractor from a business, they should give you a service book. If the thermostat is rusty and dirty, it could be sticking. To check to see if the thermostat is opening properly, you can put it in a pot with water in it and heat it up. If by the time the water boils, the thermostat hasn't opened, presto! You have found your problem.

Don't forget that tractors use the same coolant your car uses. If somebody tries to sell you some specially formulated anti-freeze for you tractor for approximately $25.00 a gallon, tell them your unicorn ate your wallet!

Hard Starting

The number one service call at our shop is "my tractor won't crank." Tractors do a lot of sitting day after day. The more often you start your tractor the less likely you'll have a tractor problem. New tractor owners need to read this part as well as used tractor owners. I've seen many new tractors sitting on a dealer's lot from 3 to 5 years before they sell.

The first thing to check is to turn on the lights. If they come on, then the battery is fine. Next you need to check the neutral safety switch. Some work by pressing the clutch pedal all the way to the floor. Others work by moving one or both gear shift levers to the neutral position. Fords are especially particular about this. Sometimes you need to turn the ignition key and wiggle the gear shift levers to make contact. When the starter begins to turn, stop wiggling.

Some of the newer models will also have a safety ignition interrupter switch on the PTO or power take-off lever. It's not a bad idea, but not everybody knows to check for it. It's always a good idea when you park your tractor with the mower still attached to turn the PTO off before shutting down the engine.

Cats and dogs like to hide under a mower, so before you lose a pet, it's best to crank your tractor, lift the mower, and make sure nothing or nobody is under it before you turn on the PTO lever.

If the lights don't burn, you need to check for power at the battery. A simple 12 volt tester works fine. If the battery is dead, then you've found your problem. If the tester shows 12+ volts, then the next thing to check would be the battery cable connectors on the battery. Take them off and clean them with a terminal cleaner or just a piece of sandpaper. If your cable connects by a clamp to the battery terminal, you need to remove them and clean them as well. Then try to start the tractor again.

If it still fails to start, remove the ground cable. This will be the cable that connects to the smaller terminal. It will have a minus sign beside it. In most cases the ground cable will be black, just follow the cable to where it bolts to the frame or engine block and unbolt it. Then sand the end of the cable and the metal where it bolts to remove any rust or corrosion.

Hard Starting

If the tractor lights burn, but the tractor still refuses to start and you're sure the neutral safety switches are working right, you can check the power with a 12 volt test light, or volt-meter, on the starter solenoid. The positive cable on the battery generally has a larger terminal and has a plus sign beside it which leads down to the starter solenoid. The solenoid is normally about the size of a cola can and is mounted on top of the starter.

At this point if you plan to continue on your quest to start your tractor without a mechanic, you must be absolutely sure the tractor transmission levers are in neutral. Be absolutely sure the parking brakes are locked.

On the end of the solenoid you will see the cable that comes down from the positive battery terminal, and you will generally see another bolt next to it. If you look between them or above these two cable bolts you will usually see a small wire that connects to the solenoid. This is the exciter wire. When you turn the ignition key switch over to start, the switch sends 12 volts of current down to this terminal, and it causes the switch inside to connect the power from the positive battery cable to the other large bolt and causes the starter to engage.

Now is a good time to remove the exciter wire from the terminal on the solenoid and ground one end of your test light and touch the other end to this wire to see if the tester light comes on. When you reach around to the key switch and turn it over to start, again be sure the transmission levers are in neutral and the parking brakes are locked. It wouldn't be a bad idea to shut the fuel lever off, just be absolutely sure the tractor won't start and run over you! If the test light lights up when you turn the ignition switch to start, then you know the problem is in the solenoid or starter.

Sometimes the solenoid switch can stick. You can dislodge it by crossing a screwdriver across the two large terminals. If the starter is stuck, this action could dislodge it also. Then you should reconnect the exciter wire on the solenoid and turn the ignition key. If it still doesn't start, the problem could be that the starter drive wedged against the teeth on the flywheel.

To dislodge the starter drive, you can put the tractor in high gear and high range. Remove the parking brake lock. Make sure the fuel stop, on a diesel, is in the off position, and then you need to rock the tractor forward and backwards. If you hear a clank sound that would be the starter drive that was wedged against the teeth on the flywheel dislodging itself. The tractor should now start by the key. If it still fails to start you need to remove the starter and take it to a garage or to a good starter-generator repair shop. This procedure could save you a $50-$100 service call.

You can thank me by telling your friends and relatives about this book.

Safety Considerations

My father said, "You can get hurt in a split second and it could take a lifetime to heal." He was right.

I have delivered several thousand tractors, and I always tell the customers how to operate their tractor. I even explain it to people who have owned one before, but especially to first time owners. I explain what everything on the tractor does and how it does it. I tell them to read all the safety warning labels before driving the tractor and only start the tractor while sitting in the operator's seat.

I always quote Bill Elliot. In an interview he was asked "How do you decide what to do when you're driving nearly 200 miles per hour and something goes wrong?" His reply should be written in stone.

He said, "You don't have time to think. You just reflex!" This was one of the best answers I ever heard in my life. I explain to people to just slowly ride around for several hours and get used to the brakes, steering, hand controls, etc. until you get used to the tractor and/or implements. After doing this you can react by reflex rather than to having to think about what to do.

I go through the normal list of bad ideas, including driving your tractor while drinking or taking prescriptions or any drugs. Never operate your tractor when you're excessively tired or under mental stress or exhaustion. Never drive your tractor fast unless you're on a paved road.

I'm a country boy. When I drive on the highway, I slow down and ease the two right side wheels off on the shoulder and let traffic by, and then I ease back on the pavement and resume speed. A person who drives a tractor down the road and holds up fifty cars is looking for trouble. Use your brain. Your tractor runs about 12-15 miles per hour. Cars run 70 miles per hour. Keep your eyes open at all times.

Never take passengers on your tractor especially if you're mowing or harrowing. Too many tragedies have taken place this way.

Try to remember: a tractor has a much higher center of gravity than a car, so you must be careful on hillsides.

If you need to tow something heavy up a steep hill, put the tractor in the lowest gear possible and let the engine idle. The extra minute or two it takes to reach the top of the hill may save your life. If for any reason the front of the tractor tries to jump up, press the clutch. It will sit back down. Only pull heavy objects from the pull bar!

Some people think, if they go up the hill on a 45 degree angle, it would be safer, but a tractor is much more likely to roll sideways than to flip from the front.

Also remember, if your tractor is equipped with a roll bar, wear the seat belt.

Safety also applies to the other people around you. Be sure not to have an audience if you are rotary mowing. A rotary mower can throw a stone, brick, or stick 50-100 yards.

Remember, before you start your tractor with the rotary mower mounted, always shut off the PTO. It is even better to shut off the PTO when you park the tractor, but be sure just in case you forgot or if someone else drove it last, be sure the PTO (Power Take Off) lever is in the off position.

Be sure before starting the engine to check the engine oil and coolant levels. This is safety for your investment.

When you start the engine, lift the mower off the ground and make sure that your kitty cats or neighbors puppies haven't moved in under the mower. A couple of times in many years of delivering equipment, a pet was lost, because we forgot to shut the mower (PTO) off before starting. Please be careful!

Every time you get off your tractor while mowing, shut off the mower, even if you're just going to open a gate or get the mail. Shut the mower off before dismounting.

One of our customers parked his tractor with the engine running and the mower turning to get his mail. After removing the mail from the box and looking through it, as he was walking, he suddenly realized his foot was under the mower just below the spinning blades. He said he froze in his tracks as he slowly pulled his foot back. The tractor and mower slowly followed. He said the first thought that hit his mind was me warning him not to dismount the tractor while mowing, unless you turn the mower (PTO) off. Apparently he had not pressed the parking brake hard enough before dismounting.

By the way, I don't trust parking brakes, wild bears, or politicians, and neither should you.

Never operate your mower when it is lifted off the tail wheel. Always disengage the mower (PTO) when transporting it to another location.

I was at a Warner Robins, Georgia, body shop in the early 80's. I saw a Toyota pickup with something protruding through the passenger door. It was protruding about 4" through the door skin toward the outside.

I asked the owner, "How did that happen?"Apparently what was left of an adjustable wrench penetrated the door like that.

He pointed inside the truck and said, "It came through that door first."

I looked inside the pickup to see an even bigger hole in the driver's side door. I turned around to face the owner.

I asked, "How did that happen?"

He said, "I hit that wrench with my riding lawn mower, son."

I couldn't believe it. I asked him again, "Are you sure it wasn't a big rotary mower instead of a riding lawn mower?"

"I'm sure," was his response.

I could only imagine how much more powerful a tractor powered rotary mower would be over that of a riding mower. So if you're mowing, don't allow spectators. It's just too dangerous.

Most rotary mowers have an option available. It's a flat bar of metal with short chains about 6" long hanging from it. The idea is to slow down objects being thrown from the mower. I offer kits at pure cost to all our customers.

Always be careful when mowing around dead trees or trees with dead limbs. Vines always seem to grow on dead limbs or trees. When you mow by them, the vines can wind up around the mower shaft or stump jumper and draw the dead limb or tree toward you. Even tractors with roll bars only offer limited protection and most roll bar tops are either fiberglass or plastic. You also have to be careful about picking up an old fence from the ground. One that has been flattened for 30 years can be picked up by a spinning set of mower blades.

If you hear something making any sort of noise when mowing, shut the mower off and the tractor also.

Now is a good time to insert this bit of advice. Almost all tractors have a tool box mounted somewhere. Most often it is under the hood or on one of the rear fenders. This tool box needs to carry extra lynch pins, hitch pins, a hammer, and a good set of pliers with a cutting edge. A few first aid items wouldn't be a bad idea either.

As I stated earlier, you need to shut the mower and tractor off, then you will need to cut the wires as they protrude from the mower. I don't suggest you try to cut them loose from under the mower in the woods or field. I suggest that after you cut the wires you should drive your tractor back to your barn or work shop to remove the wire. You may need other tools, and if it's really bad, you'll be closer to your bottle of Jack Daniels.

Next I suggest that you detach the mower from the tractor and lift it by the tail wheel with an engine hoist or come-along under a big tree limb. Then brace it as best you can. I never suggest just raising the mower with the tractor lift and working under it. It's just too dangerous.

I know of a case that happened years ago in my old hometown where a man was removing old fence wire under his mower while it was still attached to his tractor. His neighbor came over to help and saw that the lift had settled, so he started the engine to lift it back up. Unfortunately the power take off (PTO) lever was engaged and he accidentally killed his friend.

Please do not work on a mower or harrow or any other implement while it is still connected to the tractor! A tractor can be the handiest tool you ever owned, but like anything else, it can also be dangerous.

Never stand between the tractor and implements when hooking the two together. Stand to the side or better yet, do as

Safety Considerations

I have done. I acquired a pallet for each of my implements. I cut old plywood into 4x4 pieces and laid them on top of the pallets. Then when you want to hook the tractor to the attachment, just back up close to it and lift the lift arm up about even with the hitch pins and shut the tractor down. You can easily slide the implement into place and hook up the hitch. Always remember to hook up the lift arm without the adjusting link first. Then, you can turn the handle up or down to connect to the other lift arm pin.

Observe this special note. If your hitch was level before, count the turns when you lift the arm up or down, and afterwards if you turn the same lever the exact number of turns the other way, the implement will be level.

Always connect the top link, the adjustable bar, which connects the top part of the hitch to the tractor last. If you attach the top link before either one of the lower lift arms, it will drive you crazy.

Always mow new land slowly. Gear the tractor down until it moves slow enough that if you see a stump or pot hole, you can stop or go around it. Here is a helpful note. If you do wind up on top of a stump and the mower is jumping around like crazy, grab the lift lever and pick the mower up just a little and move forward or backwards off of the stump. Most mowers have a round disk the blade attaches to called a stump jumper. It's not a stump grinder, so get off of the stump quickly.

Never operate your tractor when the weather turns bad and lightning is flashing everywhere. Some people think they are safe because of the large rubber tires, but one blade of wet grass could turn you into toast. I see some people mowing in the rain. This is not a good idea. The wet grass will cake up under your mower and make a mess.

Never try to cross through a moving creek or drive through ponds with your tractor. Some people think because of the big tires it will float. I spent several afternoons winching tractors from creeks and ponds. One man said his neighbor told him tractors should float. I suggested he send him the bill after he paid us first.

I need to let you know: that nice little bumper on the front of your tractor is just to keep weeds and bushes from hitting the grill. We had one man try to fell a 10" diameter tree with it. The first attempt failed, so he figured he needed more speed. The repair bill for the bumper, grill, radiator, etc, was nearly $500.

I asked him, "Where did you get the idea that you could knock a tree down like that?"

He said, "It's a tractor, isn't it?"

This man was a college professor. I worry about his students. A tractor is a tractor; it's not a tank. It won't fly, it can't swim, and it can't read signs. It's just a tool that does lots of hard work, so you can spend more time fishing, playing golf, or sleeping in your hammock, but it's not invincible.

A man called me one Saturday morning and said he needed me to bring my roll-back truck and pick up his tractor.

I asked him, "What is wrong?"

He said, "I got some water in the fuel."

He lived on Lake Blackshear about 50 miles south of our shop. When I arrived I didn't see his tractor, so he pointed it out to me. He pointed toward the lake. That's when I noticed the muffler sticking up out of the water. I thought his neighbor's kids had filled up his fuel tank with the garden hose. I can't count the times that has happened, but his tractor had become a U-Boat.

"Water in the fuel," was my response.

He grinned and said, "I was afraid if I told you the whole story you wouldn't come."

"How did this happen?" I asked.

"I jumped off the tractor to move the lawn chairs, and when I turned around it was moving down the hill to the lake," he answered.

To help you understand better, let me explain. The air intake on a Ford 2000 is on the very top of the hood, so when the water got to the top of the air cleaner, that is when the tractor stopped.

It was evident that our customer had a passenger, a big old hound dog, along for the ride. When he got off to move the

Safety Considerations

lawn chairs, the dog that he had left sitting on the seat decided a nap was better than tractor riding, so he jumped off and must have accidentally bumped the gear shifter into first gear.

I saw the hound dog sleeping on the porch next to a trash can full of beer cans and decided not to push the subject. The owner must have been reading my mind and started laughing. I handed him the end of a long chain and said, "Laugh your way out there and hook this to the back of the mower."

So I guess the moral of this story is, if you get drunk while mowing around your lake lot, be sure to have an old hound dog to blame your accidents on.

After retrieving a tractor from a lake, try to keep your wits and remember these following helpful hints. Never try to refuel a hot tractor or refuel a tractor while the engine is running, especially a gasoline engine model. One drop of gas could drip down on the distributor, and the results could be very tragic.

As I stated earlier, be sure to never start your tractor unless you're in the driver's seat. The safety neutral start switch will normally prevent the tractor engine from starting, but over time some of them wear out and fail. Never bypass this switch. If the neutral safety switch fails, replace it. A switch is only a few dollars. How much is your life worth?

A nice young lady in our Sunday school class told us her father had been seriously injured by his tractor. She said he thought he took it out of gear while standing between the front and rear tires of a large International. He then reached over and turned the key to start the tractor. It started instantly and the rear wheel ran over him. Fortunately, he survived his serious injuries, but his recovery took months to achieve. Never start a tractor unless you are sitting in the driver's seat and have your feet on the brakes and clutch.

Never lean over the side of a tractor to try and reach something while the tractor is running. You might have accidentally left it in gear.

I see people talking on cell phones while driving a tractor. It's not a good idea. You need to pay attention to what you are doing.

When loading and unloading a tractor from a trailer you need to be very careful. Always idle the engine in the lowest gear possible. Never get in a hurry when loading or unloading. It's a good idea when mowing to wear eye protection. It is especially important to wear ear protection, if your tractor has a roll bar with a top. The top reverberates the noise and compounds it.

If a pretty girl in a halter top walks by, I suggest you stop the tractor temporarily. My brother walked off the top of a house we were building on the Georgia coast when two young ladies walked by in bikinis. Fortunately, he landed in a big sand pile.

Never reach under the hood or around the grill or cooling area with the tractor running. The fan blades or cooling vanes on the generator or alternator could seriously injure your hand. I have had people tell me they reached into the area behind the radiator to check something, thinking the radiator shroud would protect them. The radiator shroud's only function is to help the spinning fan blades pull air through the radiator. Please shut the engine off to examine anything under the hood.

Never approach the spinning PTO shaft of an implement, especially if you are wearing loose clothing or if you have long hair. A man's tie or a woman's long hair getting caught up in the spinning PTO shaft could be tragic.

Never remove factory safety shields from your tractor or implements. They were put there for a reason.

Only pull heavy objects from the drawbar. Never fasten a tow rope or chain above the drawbar area. If you hook something above the rear axle, it can cause the tractor's front end to lift off the ground.

Never hook a chain or rope to the roll bar of a tractor. The worst case scenario would be to tie a tow rope to the top of a roll bar. When you let off of the clutch, the tractor would flip over backwards. You need to explain this to your kids to make

Safety Considerations

sure they understand! Even farm kids need you to explain this. I know farm kids generally have lots of common sense, but the safety tips should be explained again and again.

Once again I want to explain don't drive a tractor fast on rough terrain. Always slow down before making a turn. If you are transporting something in a loader bucket, keep the bucket close to the ground. It lowers the center of gravity. Watch out for electrical wires over head and under the ground.

Don't operate your tractor after dark unless you have the lights on. Always operate the lights on your tractor, even in the daytime, if you are driving on the road. It is important for you to display a slow moving reflective safety sign on the back of your tractor even if you drive on a farm road. Some states require it by law. The orange triangle is an internationally known warning sign for equipment.

I recently saw a man driving an old Ferguson down the highway. He was smoking a cigarette and holding a five gallon gas can on the tractor hood ahead of him and was only wearing shorts: no shirt and no shoes. I could only imagine all the scenarios of his destruction. The cigarette in his mouth could ignite the fumes coming from the fuel can, especially after taking a drag on his cigarette with the tractor moving forward creating a draft which could excite the flame. He could drop the gas can or spill some of the gasoline on the ignition coil or spark plugs. He could be rear-ended and the can of gas could soak him completely. One spark and KA-BOOM! Did I mention the six pack of beer in his lap? This guy could be the poster child for advanced stupidity.

The older tractors like the very popular 8N and old Fergusons have what is known as a dead PTO. That means the mower you're pulling will push the tractor, if you push in the clutch, even if you stomp on the brakes.

A simple solution is an over-riding coupler. It works like a ratchet wrench when you mash the clutch pedal. This coupler just free wheels and makes a clinking noise while preventing the mower from pushing the tractor. The coupler only costs about $50.00, but it can save your grill and radiator worth approximately $500.00. And it could save you from being injured.

To sum up this section on safety, you should read the owner's manual and study the warning labels on the tractor.

Be extra careful. Take it slow and easy until you get familiar with everything on your tractor. Don't operate your tractor if you are tired, drunk, sleepy, or in an angry mood. Just use common sense. Take your time. If a situation unfolds that leads you to believe you could have an accident, just back off until you're more experienced. Regardless of how much safety someone tries to teach you, it's up to you to be careful.

Just imagine your wife and her new boyfriend spending your life insurance. Maybe that will help you concentrate!

Maintenance and Service

I know this section is a waste of time. We have sold tractors to people who traded them back 14 years later. These tractors still had the same oil filter we installed before the sale.

Then there was the guy who stormed into our shop one day with a metal rod in his hand. It was the axle bolt from the tail wheel of a mower, but it was worn down to about half the normal size in the middle.

"Did you grease this?" was my question.

"Why wasn't it greased before I bought it?" he asked.

"That was 12 years ago," I replied.

One man hauled his tractor to our lot complaining about a noisy engine. We checked the oil. It didn't even touch the stick. As dirty as the oil was, we decided to drain the oil pan rather than add oil to the engine. We only got about 2 quarts. Considering the fact there would still be about a quart in the oil filter, he was only about a gallon low.

I asked him, "When was the last time you checked the oil?"

The blank look on his face told me it had been years, if at all. Upon closer inspection we found a sharp piece of metal jammed in the bottom of the oil pan, and a slow drip was emanating from it.

I've told people until I was blue in the face, "Check your oil and coolant levels before starting your tractor."

Several times we have had people damage perfectly good engines, because they didn't check the oil level and coolant level before starting their tractor.

In the 30 plus years we have been in this business, I don't think we ever rebuilt a Ford engine because it was simply worn out. The Ford engine block, crank shaft, pistons, and rods are so big, tough, and heavy duty. This means that they can take many thousands of hours and show almost no wear. Assuming, of course that they are properly serviced and the radiator is kept clean and full of coolant.

Another important thing is, if your tractor has a vertical muffler, be sure to keep the rain cap in good working order. If the rain water gets down the exhaust and stays there for a few days, it will soak down through the valves and into the cylinders and rust them up really bad. If left too long, the pistons can rust and stick to the cylinder walls. This is the number one reason we rebuild these engines. Something that only costs about $5.00 destroys an engine worth thousands.

Some people just put a tin can over the pipe, and then one day they forget the can.

Whenever the oil starts turning dark, change it. Simple enough: just change it. You don't have to be a mechanic to change oil. It's a good idea to crank the tractor and let it run for awhile (3 to 5 minutes is good) to warm up the oil, and then shut the engine down. Remove the drain plugs located on the bottom or near the bottom of one side of the oil pan.

For most tractors you will need a 2 gallon or larger drain pan. I always remove the dip stick and oil filter cap, so the old oil will drain more thoroughly and quickly. Let it drain for about 15 minutes and then replace the oil plug. Then remove the oil filter. Before installing the new oil filter you need to pour it full of fresh motor oil, completely to the top one time. The oil will settle down as the air bubbles. This process is called wetting the filter.

Now you can install the new filter, but first make sure you smear a little film of engine oil on the face of the rubber gasket. This will prevent the gasket from pinching as it tightens. Also be sure the old rubber seal didn't remain on the engine block. Be sure the old oil seal ring is on the old filter you removed. Spin on hand tight. I like to give it about a one-third to a one-half turn with a filter wrench. I know the oil filter instructions say hand-tighten only, but I would rather be safe than sorry. Besides, don't the same people that sell oil filters also sell engine parts?

Next wipe off the dip stick and replace it. Now pour in the new engine oil. It's always a good idea to start the engine afterwards for a couple of minutes. Just let it idle. After the oil gauge starts to register or the idiot light goes out, shut the engine off and wait a few minutes. Pull the oil dip stick out and wipe it clean. Re-insert the dip stick and check to see if it's full. If you're not sure how many quarts your tractor holds, call your dealer. Then take a magic marker and write it down

Maintenance and Service

under the hood. If you're like me and have a bad memory, also write down the size wrench required to turn the oil plug.

The next step is to get your handy grease gun down and grease the fittings on your tractor and implements. Most tractors have grease fittings on the front wheel hubs and also on the back of the steering spindles on each side. Some have fittings in the center of the front axle near the center where the axle connects to the front mount. Be sure to look under the pedals on both sides.

The tie end rods would be the next place to check. Many of the tie end rods today have no grease fittings. Ahhhh, that planned obsolescence.

Next, check the rear of your tractor including the adjusting arms which raise and lower the lift arm. Check your owner's manual to be sure you have everything covered.

Now is a good time to grease the implements, so that you don't forget them. Remember, about three pumps on the grease gun per fitting should be sufficient. In some cases, for instance, wheel seals, these seals can be damaged by pumping too much grease into them.

Oh, by the way, just standard axle grease is fine for most farm tractors. The more expensive lubricants generally just cost more.

While you are going over your check list, now is a good time to check your coolant level. Remember, a 50/50 mixture with water is just fine. Don't fall for the $25.00 a gallon anti-freeze sold on some websites. They claim you have to use it in a farm tractor. Phooey!

Don't fill the radiator all the way to the top. Leave about 1½ inches from the top for expansion. Next you need to be sure the radiator is clean. When your engine is running, the fan blades behind the radiator pull air through your radiator. Unfortunately the air will be full of dust, leaves, and grass, so be sure to either blow the trash off the radiator or wash it out with a water hose.

I never like to put cold water in a hot radiator. Let it cool first. Yeah, I know radiator shops say it won't hurt it, but they also sell radiators, don't they? When your radiator is clean, you should be able to see through it, so remove the grill and clean it well. P.S., be sure the engine isn't running while cleaning the radiator.

You also need to check the hoses for cracks and the radiator hose clamps for tightness. I would also make sure the fan belt is tensioned about right. You should be able to flex the belt in between the crank shaft pulley and the generator or alternator about ½ to ¾ of an inch.

If you find the coolant is low a couple of times in a row, you need to inspect the water pump to see if coolant is dripping from the center of the water pump. For novices, the fan blades bolt onto the face of the water pump.

Now it's time to check the air filter. If you don't know where it is, you can check the owner's manual or look to see where the air intake hose goes. Maybe you'll get lucky. Some older tractors have an oil bath air cleaner. Some are located between the grill and the radiator. Still others will be on the side of the hood or cowling. Loosen the clamp around the bottom and remove the bottom oil cup. Wash it out clean with gas or starting fluid. Fill to full marks and re-install.

For dry element filters, just remove and replace. If your tractor dealer is closed on Saturday or a long distance away, go to a NAPA dealer or CarQuest, they can usually match up the right filter for you. Be sure to take a damp cloth and remove any dust or debris left behind in the air filter container before you re-install the new filter. It's a good idea to take a magic marker and write the date on the filter for future reference. Then write down the air filter number under the hood.

You've done really well so far, but don't get cocky on me; we still have a few more things to check before you get your diploma.

It's time to check your transmission lube. On some tractors the transmission lube and the rear end lube run together. This is a good time to consult your manual. If the transmission and rear end have separate compartments, you will have a drain plug under each one. When you drain the transmission, it will usually use 85W90 weight gear oil, but

Maintenance and Service

be sure to check your owner's manual. This heavy lube needs a long time to drain. I would remove the drain plug and also remove the fill plug which should be on top of the transmission housing. This will let the lube flow more freely. I would give it several hours to drain completely. If you used one drain pan to drain the transmission for about 15 minutes, then moved that pan and replaced it with another pan and let it drain for a couple of hours, you would be amazed at the sludge and nasty looking gunk you would have in the second pan. Most people get in a big hurry and leave this mess in the transmission. Be patient. Have a Pepsi and relax for a while. You've earned it.

Break's over! Back to work. Now place a pan under the rear end if it's separate, and drain it.

After re-filling the transmission and rear end with the proper lube, run the tractor for a few minutes. It's OK to just drive around the field and come back. Then let the tractor sit for at least an hour. Afterward, double-check the lube levels.

Special note: Be sure to check before you buy the hydraulic fluid for the rear end. If your tractor doesn't require non-foaming oil ($100.00 higher per five gallon bucket), don't waste the money. If you have money to give away, you can mail it to us at 2191 Hwy 247C, Byron, GA. 31008.

Now is a good time to spray a little lubricant on the hood hinges and latches, lift arm balls, PTO splines, and battery terminals.

When performing any task, it's important to clean up your mess and recycle the used oil and lubricants. Most auto parts stores and tractor dealers will accept your old oil and give it to a recycler.

Now it's time to change the fuel filter. If it's a gas tractor, you could have a glass bowl separator. Just turn off the gas at the shut off under the fuel tank. Then remove the glass bowl and wash it clean. Re-install and turn on the fuel. If it's an in-line fuel filter, shut the fuel off and just replace the filter. Be sure to install the filter with the flow arrow in the right direction.

If your tractor is a diesel, shut the fuel off just like you did with the gasoline engine tractor. If your tractor has a spin-on diesel fuel filter, just spin the old filter off. You will need enough fresh clean diesel fuel to completely fill the new filter until all the air works it's way out. I just sit the new filter upright in the filter box and use the box as a holder. Then screw the new filter on being careful not to spill the fuel out of the filter.

Next turn the fuel on and start the engine. If your tractor has a cartridge type filter, turn the fuel off and remove the center bolt that holds the filter on. In most cases, Ford and Massey in particular, the filter will come with 3 gaskets: 2 large ones and a smaller one. The larger gaskets go on top. Remove the old gaskets from the bottom cap and the top mount, and then look up inside the middle of the top. You will see the smaller O-ring. Remove it and install the new gasket.

Then, install the two large O-rings on top and bottom of the new filter. Hold the entire assembly together and then reach up on top with the other hand and turn the center bolt clockwise by hand. Then hold the filter assembly as tightly together as possible and tighten the center bolt with a 7/16 wrench until it is snug. Be careful. The assembly is made out of pewter (junk metal), so don't over tighten the bolt. Then look on top of the assembly until you see a 9/16 bolt. This is a bleed screw to let the air out of the filter. Loosen it several turns and then turn on the fuel at the base of the tank.

Special note: The fuel tank needs to have at least five gallons to flow properly through the new filter. More men have wound up in hell trying to bleed the fuel system with a near empty tank, while trying to crank a lawn mower, than you can count!

It should only take about a minute for the fuel to reach to the castellated bleed screw. This bleed screw has a lengthwise groove in the threads to let the air escape, but seals completely when tightened. First you will see air bubbles and fuel, and then straight fuel. Now is the time to tighten the 9/16 bleed screw. Be sure the dash mounted fuel shut off knob is pushed in and the throttle is about half way open.

On a Ford you need to open the hood and prop it open with

the hood prop assembly. You will see a bolt on the right hand front corner of the battery tray. Loosen the nut enough to spin the battery tray out. It will pivot on the left side. Then take a 5/8 wrench and loosen the cap on top of the injector. Just loosen about a turn, don't remove these nuts. Be sure the tractor is in neutral and reach back around with your left hand and engage the starter on the key switch.

After the engine spins over a few times, you should see air bubbles and diesel fuel spitting from the loosened cap nut. When you see no more air bubbles on one of the injectors stop, then tighten it with your 5/8 wrench. Repeat the process until the diesel fuel line nut is tight and the engine is running. Let the engine run for a few minutes and check for leaks. Presto, you're done!

If you have a hand pump on the side of the injection pump, you can pump the fuel up to the injectors with it rather than using the starter.

You have now become a tractor service man and earned your diploma. Pat yourself on the back and clean up your mess.

Protect Your Tractor From Weather

I always suggest you park your tractor under a shelter or in a barn. Tractors left out in the weather generally have more up-keep problems. Rain water will seep into housings and contaminate the lubricants. The sun causes wear and tear on hoses, wires, seats, and fades the dash gauges. It can also cause the shift levers and steering wheel to age and crack prematurely.

Be absolutely sure, if you leave your tractor outdoors to cover the open end of a vertical muffler. If your tractor has a vertical muffler, you need to pay close attention to the rain cap or flapper on top of the pipe. If it has any damage, you need to replace it. Most of the tractors that we have rebuilt the engines in over the past thirty years had to be rebuilt because rain water got down inside the muffler, then dripped into the engine and rusted the cylinder walls and valves. A rain cap only costs about $5-$10 dollars, but an engine overhaul can run several thousand dollars. I tell people the rain cap is the second most important part on the tractor, only the driver is more important.

Some tractors with vertical exhausts will have an exhaust pipe on top that makes a 90 degree turn and throws the exhaust out the side. These are alright. But, be sure that the part of the pipe that bends has not corroded or rusted to the point that it has holes in it.

It's a good idea when you plan to park your tractor during the winter to fill the fuel tank. My experience shows a full fuel tank won't have as much condensation form during the time it's stored.

I suggest that for storing your tractor during the winter that you remove the battery cable. Rodents sometime chew on the wires and can set your tractor on fire. If it's in a barn, you could lose it, too.

I put ground disconnect switches on my farm trucks and on other equipment that I leave for days at a time. When I leave them, I just throw a switch and it cuts the power off to the electrical systems by disconnecting the ground. This not only protects your equipment from accidental fires, but could prevent a thief from stealing it. On older tractors, especially the ones 20 years or older, the wires are beginning to age and dry crack and could develop a power drain. This is a good way to be sure your battery will be strong when you return. These ground disconnect switches are also a blessing if for some reason your equipment develops an electrical fire. You can quickly shut off the power by just flipping a switch.

It is also important when storing a tractor or other equipment with a manual transmission to press the clutch down and block it in that position. A piece of wood would work fine. This will disengage the clutch from the flywheel and prevent it from sticking. If it sticks and you don't know it, you could crank your tractor and possibly run into something, because depressing the clutch pedal wouldn't stop the tractor. If this happens to you, shut the engine off immediately and press hard on the brakes until the tractor stops.

Tractors and equipment left in flood water for only a few minutes or in a continuous rain could wind up with a stuck clutch. In 1994 we experienced our first flood in Middle Georgia. A tropical storm left over from a hurricane parked over our area for about a week. Many people were caught off guard. We had never experienced such a thing before.

I, by the way, would like to apologize for this event. I believe I caused the flood, because I was heavily considering getting into the irrigation business.

My phone started ringing. People were telling us that their clutches were stuck and one man said he accidentally drove his tractor through a new board fence.

I called the local television station and asked them to warn people about this situation. None of the television stations gave out the warning that I know of. I called some radio stations, and they warned people. I guess that the television stations were run by city slickers, while the radio stations must have had some country folks running them.

Good Tips Before Starting

I tell everybody, especially when your tractor has been sitting for several days or longer, to check oil, coolant, and all of your lubes (transmission, rear end, power steering, and hydraulic system, if you have a loader) to be sure they are full. In some cases people will strike something while out mowing or plowing that could puncture a hole in the radiator, oil pan, or one of the other systems, such as a cooling hose or hydraulic line, without knowing it. It's always best to check them to be sure.

Tractor Brands

For the rest of this book, we are going to discuss the most important brands of tractors you will run across. We've tried to include most of the important ones, and talk about different variations to look out for. But, there's just too many brands and models on the market to talk about all of them. Stick with the good brands and models like the specific Fords we talk about next, and you will have a lot fewer problems.

Of course, this book is just for fun, so don't take any of our advice seriously. If it works out great, we'll take the credit. If not, use this book to start a fire.

Ford

Ford tractors are my favorite tractors. We have sold them on our tractor lot for over thirty years. When I was 6 years old, I came home from school on the bus and saw an 8N Ford sitting in our front yard, and it was love at first sight. We already had a C Allis Chalmers my father had bought new before I was born, but he wanted a Ford. This was a quality tractor. Apparently Henry Ford had admired the TO-20 Ferguson. It was low profile, wide front end, small, and easy to handle. It was the first modern style tractor. Mr. Ford just improved the Ferguson. Even though they look just alike, they are completely different. As far as I know, only the tires and battery will interchange.

The Ford 4 cylinder engine is a much better engine than the Continental. I couldn't count the number of Fergusons with busted blocks. Many people just bored holes and put bolts through the block after they cracked and filled them full of block sealer.

It's hard to find a good 8N today, but there are still plenty of them around. I still have the one Dad bought in 1957. It still looks and runs as good as it did over fifty years ago. Most people who own them wouldn't sell them for love or money. Now even though the 8N was light years ahead of the competition, it did have one handicap, the dead lift. Now what I mean by *dead lift* is that the hydraulic lift system doesn't lift unless the clutch pedal is up. With a dead lift, if you stop at the end of a row, you have to take the tractor out of gear to lift the plow. A *live lift* means that the lift will operate anytime the engine is running regardless of whether the clutch is depressed or not.

Now, real farmers with a dead lift learned to raise the lift handle just before the end of the row. As you know, the PTO (power take off) is the splined shaft on the back of a tractor that turns the drive shaft on a mower. With these models, the PTO has to be turning to operate the hydraulic pump because the hydraulic pump is located in the bottom of the rear end. This is called a *dead shaft*. A *live shaft* means that the PTO turns anytime the engine is running and the PTO lever is in the ON position.

I love these old 8N's, but I personally think they are too old and difficult to use for an inexperienced operator.

The next tractors, the NAA and Jubilee, have a live lift, but still have a dead shaft. These were made in 1953 and 1954. This model also left the old 4 cylinder flat head engine behind. If you can find one of these, you will notice they still have a large nut on the end of the rear axles.

In 1955 Ford came out with the 600 and 800 series tractors. The rear axle nuts disappeared. If you think about buying an 8N, NAA, or Jubilee, be sure to check the rear wheel axle nuts. If you can grab the top of the rear tire and get it to wobble, the repair could call for the axles to be replaced. Or, some people weld them with a couple of spot welds. I would just avoid them, if they are bad.

With the advent of the 800 series, Ford offered two tractor sizes and a diesel engine. They also offered a 5 speed transmission with a live PTO shaft. They were trying to keep up with Massey Ferguson's 35. Unless you can find a rare 600 or 800, I think they have been around the block or field a few too many times.

In the early sixties Ford started painting their tractors blue and gray instead of red and gray and called the 600 and 800 series tractors 2000's through 4000's. In 1965 Ford restyled their tractor line. The tractors were given square shaped hoods and new engines. The 2000, 3000, and 4000 had a three cylinder engine instead of a four cylinder. They also offered more sizes. They now made 2000, 3000, 4000, 5000, 6000, 7000, 8000, and 9000. These tractors ranged from under 40 to over 100 horsepower. These tractors were built until the mid seventies and are some of the best tractors in the world.

I have sold a few thousand of these 2000, 3000, and 4000 tractors over the years. They are very reliable and the parts are very reasonable. The only difference between a 2000 and 3000 is the crankshaft, hydraulic lift capacity, and factory original tire sizes. The average person thinks a 3000 is a lot bigger than a 2000. WRONG! It's in your head. The only way to tell them apart is to look for the spring to be visible behind the seat of a 2000 and covered on a 3000. This spring in question is the draft control shock suppressor. If it is covered, it's a 3000. If it is visible, it's a 2000. Now, you are educated!

I'm going to mostly deal with the smaller tractors, because these are the ones that are so popular with people who move to the country and buy a few acres and realize they need a tractor.

From the time I started my tractor business in 1979 these Ford tractors have been my biggest sellers. Fords are rugged and dependable and easy to work on. Parts are readily available. Many interchange with parts from the Ford pickup of that time frame. A Ford is easier to sell, and if you get tired of it, you can generally get your money back if you take care of it. When new, these 2000, 3000, and 4000's sold from $3000-$4000. When we remanufacture them, they now sell for double what they cost new. If you buy a good quality used tractor at a fair price and keep it in good condition over time, you could make a profit on it.

These tractors have several transmissions available, a 4 speed, a 6 speed, an 8 speed, and a 10 speed automatic. The standard 4 speed only has one shift lever and no high-low range. The six speed is a 3 speed with high-low range. Then you have, my favorite, the 8 speed, 4 speeds plus high-low range.

Then you have a cursed transmission. These tractors have a ten speed, automatic with the tee shaped control handle that protrudes toward the driver from the dash just below the steering wheel. I think the worst thing wrong with the ten speed was the original drivers. They shifted them from forward to reverse without stopping and most of them were destroyed a long time ago. I still occasionally find one that works perfectly. In these few cases it's been owned by someone who took care of it. Unless you know equipment very well, I would avoid the ten speed. These tractors came with gasoline and diesel engines. You could get many options like a roll bar, spin-out-wheels, and power steering.

The older Fords steered nicely, but I believe when Ford saw Massey Ferguson selling power steering for $500.00 (this probably doubled the profits on the tractor) a strange thing happened. I started noticing some Fords were harder to steer. It only takes three revolutions of the steering wheel, lock to lock. The older ones took four revolutions. I'm one of those conspiracy people who believe this was done to coerce people to buy power steering. I've only found a few of these. The more turns lock to lock, the easier it was to steer.

The engine block, crank shafts, transmissions, and rear ends are built like a tank. The 4000 and larger versions offered a larger rear end, and with a live PTO shaft and wet brakes. Wet brakes last longer and will actually stop the tractor even going down hill. In today's market I don't have faith in anything built now like I have in those old Fords. It is still possible, even though they started making them over 40 years ago, to find one in excellent original condition.

Now I need to warn you about Ford tractors 2000-4000 series that were sold new in Great Britain and brought to the USA used. Tractors in Europe will usually have up to ten times the hours and wear. In most of Europe they will drive tractors many miles to market pulling several trailers. Most farms over there have only one tractor, and the owner drives it everywhere. I have seen some of these tractors with over 10,000 hours. Most of the ones sold here being the same year model would only have 2000-3000 hours.

Also the weather over there is much like the weather in Canada, cold and wet.

The way to spot these tractors is to check the serial numbers. The serial number on the Ford tractor is stamped on top of the bell housing just in front of your right foot, if you were sitting on the tractor. The models built and sold here have a "C" stamped in front of six numbers. If the tractor has a "B,", then it was built in Great Britain.

When our trade dollars buy more in Europe, people ship their tractors to the USA. My experience has been that these tractors have a lot more wear and tear and rust. If I find the "B" before the six numbers pressed into the housing, then I usually don't buy them at all. I've had some that were in such worn out condition that I just dumped them on the salvage row at an auction. My experience with used tractors from

Europe, even though some of them are identical to ours, has been all bad.

If I'm estimating the value of a Ford 2000-4000 that has the "B", I cut the value in half.

These tractors also have lights in the grill, but some dealers remove them and bolt lights on the side in order to fool people. They have brake lights and turn signals, because remember I said that in Europe they are driven like a car. If you see a place for a license plate, that should be a dead give-away. In some cases the rear fenders are strange looking. Many of them once sported cabs with heat only. Remember what I said about the weather being cold, like in Canada. Another dead give-away is to look for a foot throttle just to the right of the brake pedals. You need to look close. Some dealers remove all of the above items and "Bondo" over the "B" in the serial number. Your pocket knife could help you find the truth.

Some used Ford tractors are imported from Europe and may have an "A" in front of the serial numbers. These generally aren't as sorry as the ones with "B", but I'm still pessimistic about putting money into them. Another give-away is that the 2000-4000 Ford tractors built and sold in North America have serial numbers beginning with 1, 2, 3 or 4. Most often the European models begin with an 8 or 9. This again will warn you to be careful.

My experience has been that these tractors have tremendous wear and very little service. Before I learned the hard way not to buy these units, I lost money on several of them. I don't think they are worth re-building.

Ford imported a tractor called a Dexter and later a version called the Super Dexter from England. They also imported the 5000 Major and 6000 Commander about the time they were selling the 2000-9000 series. These tractors did not hold up like the regular series Fords, and I have a poor opinion of them. These tractors were plagued with transmission and lift problems. Even though the regular Ford dealers sold them and parts were readily available, they were generally much more expensive, just like the import of today. If parts have to be shipped across the world and stored, they are generally going to cost more.

As I said earlier, I don't have a good opinion of the import Fords, and I wouldn't buy one for myself. If I get one on trade, I send it to an auction. Normally, I don't give much trade-in on them, so if it sells for salvage value, I will generally get my trade-in money back.

I know if you look around, you will find a few diehards who think a Dexter is just as good as a Ford 2000. I know a guy who still owns a Yugo; it doesn't run, but he still likes it! Many of these imported tractors had Perkins engines and Belgian-made transmissions. To me, they are more like a Massey Ferguson than a Ford.

The 2000-9000 series sold so well that Ford couldn't build them fast enough to keep up with the demand, so they imported the Dexters, Majors, and Commanders from England. In the early 1970's the EPA and the unions had made manufacturing so expensive in the USA that Ford moved its North American operations to Canada. Ford started selling a small tractor called the 1000. It was imported from Japan from a company call Shibaura. It is a 2 cylinder diesel and vibrates like crazy. The parts are expensive, and it's not as durable in my opinion.

In 1975 Ford updated the 2000-9000 series by tacking a 600 on the end. Now they would be called a 2600-3600-4600 etc. The grills were changed and the decals were different. They now sported alternators instead of generators, but the only other improvement I liked was the spin-on hydraulic filter mounted on the hydraulic pumps on the 2600 and 3600 models.

Before this was done it was a pain in the rear to service the lift system on the 2000 and 3000 Fords. They didn't have hydraulic filters. Instead they used a catch screen that had to be removed on demand. It could take up to two hours just to clean the screen. You had to remove the seat, drain the lube, undo the lift arms, and then remove the top plate of the lift. Then, you had to remove the PTO shaft lever and an inspection plate on the left side.

Now do you know why I like the 2600 and 3600 series? You could spin off one filter and spin on another one in under two minutes.

I give the 600 series tractors an A+ for quality. They usually sell on average for $1000-$1500 higher than the 1000 series. These tractors are the cream of the crop. I have probably sold a thousand of them with nearly 100% satisfaction from our customers. We don't sell anything we don't have faith in.

My father told me over thirty years ago when I started this business, "Don't sell anyone a tractor you wouldn't sell me." So if a tractor doesn't impress me, I don't sell it to the public.

Most of the dealers, who sell those weird import tractors with names you never heard of, don't use them on their personal farms. Almost all tractor dealers have farms, and almost none of them who sell these weird foreign brands use them on their own property. If they don't trust them, then they shouldn't sell them.

Once I asked a dealer who sold these imports, "Why do you have a John Deere and two Fords on your farm? Why don't you have an I-M-T tractor in use?"

He said, "You sell them, but if you're smart, you don't bring one home."

I have faith in Fords, and I've noticed even dealers who sell other brands in most cases use a Ford industrial loader tractor. Most auctions use a Ford industrial loader tractor. They just hold up better under heavy use.

I will take this time now to mention the industrial versions of Ford farm tractors. These tractors have stronger front axles, heavier built steering systems, thicker steel wheels, heavier ply tires, larger power steering systems, larger water pumps, bigger radiators, bigger brakes, etc. Ford made two heavy duty versions of the 3000 called the 3400 and 3500. The 3400 has an agriculture type grill and the 3500 has a heavy steel grill.

You have the same set up in the industrial versions of the 4000 Fords called the 4400 and the 4500. Both have all the heavy duty components, but the 4400 has an agriculture grill and the 4500 has a heavy duty steel grill. If you need to move large bales of hay or large amounts of clay or rock, you need to buy one of these industrial versions.

Some people just put one of those light-weight, quick disconnect loaders on a standard agriculture tractor and end up with a sagged or collapsed front end. If you work in construction work, you need the heavy duty industrial version. These units have heavy duty loaders that are self contained. They have their own hydraulic pump powered by a shaft mounted to the front crank shaft pulley and a built in hydraulic reservoir, so the system is completely separate from the tractor's hydraulic lift system. This means you don't have to stop and swap the hydraulic levers back and forth between the loader and the three point hitch system while you are operating the tractor.

Trust me, it's a pain in the rear when you operate most quick disconnect loaders. For these, you have to keep changing the remote control lever between on and off.

Newer versions of the 3400 and 3500 are the 340, 340-A, 340-B, 340-C, and 340-D industrial model numbers for these units. 420, 445, 455A, 455C, 455D, 535, 540, 540A, 540C, 540D, 550-555, 555A, 555B, and on are the industrial numbers for the 4000-4600-4610 etc. versions. These are the same units that also have factory mounted backhoe units.

Ford backhoes are by far the most dependable backhoes in the world. They are less likely to break down, parts are cheaper in comparison to other brands, and you don't have to be a rocket scientist to work on one.

If you have a job that requires a heavy duty tractor, I have tried them all: John Deere, International, Case, Massey Ferguson, etc. I'll put my money on a Ford any day of the week. You can run the heck out of them, just keep them serviced. After you have owned one long enough, you can sell it for a profit.

You can tear down any other brand of tractor and find that a Ford has bigger gears, a larger crank shaft, and thicker steel housings. Fords are so rugged and dependable that you almost never see one junked. Tractors are junked when they get to the point that it would cost more money to fix it than it would be worth. Because Ford parts are so reasonable and the running gears are so sturdy, you will almost never see one junked or scrapped for salvage.

We have bought Fords burned to a crisp and re-built them and made a profit. Ford is the only brand for which you can expect the integrity of the housings and internal gears to survive a fire, and then be able to buy all the necessary parts to put it in running shape at a reasonable enough price to sell for a profit.

The quality is reflected in the re-sale price of a used Ford. You can generally buy any brand on earth cheaper than a Ford, but you get what you pay for.

In 1981 Ford tractor assembly plants were moved to England. The tractors were the same quality as the units built in the USA and Canada; but because of the EPA, OSHA, and the unions, Ford just transferred the manufacturing process to England to save money. I must admit I was concerned. I still remember the Dexter, Majors, and Commanders Ford imported from England back in the seventies.

Thankfully, the 10 series tractors, 2610, 3610, 4610, 5610, etc. still had the same North American quality of engineering as the 1000 series and 600 series tractors. I don't think Americans would have accepted anything less.

By this time Ford was importing several small compact tractors from Shibaura in Japan. The 1000 Ford series were being replaced by the 1100-1300-1500-1700 and 1900's. These small tractors were built to compete with Kubota, who by now had made a real hit with the suburban homeowner who owned an acre or more of land.

I don't have anywhere near the faith in these tractors as I do in the real authentic Fords.

These import units had 3 cylinder engines that were a lot smaller and the quality had improved, but if you want to do something besides cut grass, I'd stick with the real Fords. Again, I don't like the much higher-priced parts for the compact, I must admit they were reliable when used for the purpose they were intended, such as light work, mowing, and pulling a small scrape blade or trailer around your property.

The 10-series tractors were equipped with a new grill made of metal instead of plastic. I definitely liked the new metal grill, but the top of the grill was covered by a fiberglass cover. I would have preferred metal. The hood, grill, and decals were the major differences between the 600 series and the 10 series. A couple of years later they came out with a 2810 and a 2910 with only a few more or less horsepower. I don't think a rocket scientist could tell the difference between them as far as power. But, by then all of the standard Ford tractors had the heavier wet brakes, which, until then, you had to buy the 4000 Ford or larger to get this feature. Those models had the live PTO shaft when, before then, had to be ordered as a special option.

I was happily relieved to see that the 2610 and up Fords were the same quality as before. If I were looking for a 40+ horsepower tractor, this is the last pure Ford. The smaller compact Fords were still being built by Shibaura, and they had the same new grill as their bigger counterparts.

If I had to rate the 10 series compact Ford tractors on a scale of 1-10, I would give them a 7. The fact that Ford had an outstanding parts supply system throughout its chain of dealers meant a lot, and they were as reliable as most imported compact tractors.

For a long time Ford tractors continued to be the "big dog on the block." Almost everybody in the tractor business would use Ford as a gauge to compare quality and rugged dependability, but trouble was on the horizon.

In 1990 Ford slightly re-tooled their tractors and came out with the 30 series. The lightweight agricultural front ends on the smaller tractors were gone. All of them, except the compacts, now had the heavier front ends like the 4000 and up tractors of the past. Now you had a 3230 instead of a 2810, a 3430 instead of a 2910, and a 3930 instead of a 3910. I must admit, I was glad to see them shed the lightweight front axle system. Too many people put front end loaders on them only to have them sag or collapse the front end. The power steering system was re-vamped and the top part over the grill had been re-shaped, but the "30" series tractors passed my test for strength and durability. I was satisfied that Ford tractors would continue making quality products and I was happy with them.

One thing I have found is when you least expect it, strange things happen. One night in the early 1990's I got a phone call from a good friend. He told me that Ford Tractor Company had been bought out by Fiat. I couldn't believe it. In my opinion Fiat tractors, called Long tractors in the USA, are no where nearly as good as Ford tractors.

I had hoped he was wrong, but unfortunately he was right. Then I held out hope that they would drop the Longs and run with the Fords, but to my dismay Fiat planned to drop the world's best tractor (Ford, in my opinion) and replace it with the Fiat or Long.

I had owned a few Longs and they were hard to operate. They had one gear shift that handled both the high range and the low range gears. You had to twist and turn the shifter to get the right gear. I hated to drive one. They were, in my opinion, poorly designed, and their resale value was in the basement. The poor resale value proved to me that the American public didn't care much for them either. I also thought they were the worst tractors to service or repair.

I had a friend who bought a new Ford backhoe in 1975. He installed septic tanks and did some lot clearing. The next year he went back to the local Ford dealer to get a new tractor to help with the leveling and grading. He let the dealer talk him into buying a Long in order to save a few hundred dollars. He told me later that he hated that Long and deeply regretted buying it. He tried to sell it when it was still almost new, but he couldn't get even half his money back. He eventually parked the Long and bought a Ford.

He said, "Buying that Long was a costly mistake. I thought I was going to save money, but I lost money instead."

You can get just about any garage in America to work on a Ford tractor. They are so simple to service or repair and parts are readily available.

Ford

On the other hand Long is probably the hardest tractor to find someone who will do service and repairs. Most mechanics don't want to work on something that strange or something that they might have trouble finding parts for.

I think Fiat tried their best to brainwash the American people into thinking a New Holland was a Ford, but the people who know equipment aren't so easy to fool. They sponsored a NASCAR driver for several years. I believe it was an attempt to brainwash the public into thinking Ford and New Holland was the same thing. They even painted some of the Longs to look like Fords and called them 3010 instead or 3910 and 5010 instead of 5610, but I think most people knew better.

The deal between Ford and Fiat was that they would put Ford and New Holland on the tractors until, I believe, the year 2000, and then afterwards the Ford logos would have to be replaced with the New Holland logos.

I believe Henry Ford must have rolled over in his grave.

I get calls from city councilmen and county commissioners wanting to know why the New Holland tractors don't hold up like the original Fords. I give them my opinion about Fiat (New Holland). I'm surprised at how many people still don't realize that a New Holland is not a Ford. So now we have Fiat tractors called New Hollands.

John Deere did something similar when they bought Zetors and painted them green. I think it's wrong to do something like that, but corporations get by with things that you and I would be put in jail for.

Several years ago a county commissioner from a neighboring county called and asked me if I would rather have two new Fords or three new Zetors. I told him I would rather have one new Ford over a dozen Zetors.

He asked, "Why?"

I said, "The Ford would still be running 50 years from now, but the Zetors would be broken down in a year or two."

He just laughed and said, "Thank you for your opinion."

One day on the way to my tractor shop, I saw a Zetor tractor, used by the county to pull a mower, broken down at the corner of Lakeview Drive and Houser Mill Road. It sat there for a couple of years. I later saw two more of them with the same county seal on the side broken down in two other locations in the county. I guess they got the message. The county learned a hard lesson.

A couple of years later, I saw the same Zetor type tractors painted green with John Deere decals being sold along side the regular John Deere tractors. I asked a John Deere dealer about this situation.

The dealer said, "I was told over time John Deere would make improvements in the Zetors to bring them up to the John Deere quality."

Ford has such a good reputation, except for the Dexters, Majors, and Commanders, that some dishonest people will paint another brand of tractor Ford blue and try to convince you that it's a Ford.

A man came by my tractor lot one day and said he needed a water pump for a 3600 Ford. I pulled one off the shelf and laid it on the counter.

He immediately yelled, "That's not it. That's the same one the Ford dealer tried to sell me!"

I told him, "If this water pump won't fit your tractor, then it's not a Ford."

He left with a red face trying to tell me I was wrong. Later that day he called me back and asked me if I could help him figure out what kind of water pump he needed, if he hauled the tractor to my business.

On another occasion I was showing a man a backhoe I had for sale when I looked up to see a pickup pulling in our driveway with a Long 360 on a trailer. The Long 360 was painted Ford blue and had 3600 Ford decals, but it was still a Long 360. I tried to tell the man he had a Long instead of a Ford.

The man yelled, "Can't you read. What in the h--- does that say?" as he pointed to the decal on the hood.

I showed him what a real 3600 Ford looked like and explained to him he had been cheated.

He explained that he had bought the tractor and mower at an auction only a few days before. If you buy something at an auction, you need to know what you are doing. Amateurs should never buy from auctions.

I have been in the equipment business for over 30 years, and I still get burned at auctions. Auctions don't guarantee brand, name, model, style, or anything else. They are just acting as an agent or salesman. It's not their job to verify everything they sell. The sales ticket or receipt from them has just such a disclaimer, in most cases.

This poor man I referred to earlier had paid Ford money for a Long tractor.

If you buy a New Holland tractor, that is exactly what you will be doing. I'd rather have 1 good used "20-30 year old Ford" than 10 new "New Hollands."

Fortunately for us Americans there are hundreds of thousands of used Fords around, and I honestly believe, if you take care of them by keeping them serviced and by not loaning them to somebody named "Bubba", they will last another 50-100 years.

I still have a 1951 8N Ford my Dad bought in 1957. It still runs and looks just as good now as it did then. I plan to leave it to my kids, and they can leave it to theirs.

It is important today before you buy any new or used tractor to do some research and find out what you are really buying. If you buy from a dealer or individual, make sure they put in writing what kind of tractor you are buying. Never accept a receipt that simply states "diesel tractor!" If you want the age of a Ford look right behind the serial number which is stamped on the right side of the bell housing just above the accessory mounting holes.

I just walked over to a 2810 Ford sitting on my lot and got the code. It said 3-C-16-B. The first number is the last digit of the year in which the tractor was made. In this case Ford didn't build a 2810 in the seventies or nineties, so this tractor is a 1983 year model. The letter C signifies the month the tractor was made, such as A for January, B for February, C for March, etc. So now we know the bell housing was made

March 1983. The number 16 tells the day of the month, so you know this tractor bell housing was made March 16, 1983. This should put a stop to the people who paint a tractor and decide to sell it as one that is ten years newer. For this reason, pay special attention to the code.

Over 99% of the tractors we have sold have been used. We have on occasion sold new ones for area dealers who were trying to sell enough units to keep their franchise, but on the whole we mostly sell used tractors. We put up to a year warranty on some models. These are the ones we completely rebuild, and we service what we sell. We don't mistreat our customers, and we have more repeat business than anybody else in this business that I know of. We are members of the Better Business Bureau, and we appreciate your business. We refuse to sell something we wouldn't own, so we get some trade-ins that go straight to an auction or a salvage parts dealer.

I must also warn you about a tractor called a Farmtrac. In my opinion, it's no better than the standard Long. As a matter of fact, I believe it is worse. My advice for a potential Farmtrac buyer is as follows, "It may look like a Ford, smell like a Ford, and have some Ford-like characteristics, but it's not a Ford. Isn't it strange how so many people want to imitate a Ford, but a Farmtrac is not even close?"

I have had people come by my shop who bought one of these tractors, and they try to sell them to me for half price, when they have only owned them for a few days.

I'm wondering what Fiat might have done to Chrysler. I'm still holding on to my Confederate money. If things don't improve, it will be worth more than our dollar. Regardless, Confederate money is worth more than one of those Ford knock-offs, no matter who makes them.

I have been in the farm equipment business for over 30 years and my professional advice is, "there is no substitute for Ford quality."

Henry Ford knew that!

Massey Ferguson

I first started dealing with used farm tractors around 1976. I married the most beautiful little girl that year, and we watched my house-building business crash and burn under the Carter years. My father gave me a suggestion.

He said, "Why don't you go out and buy some good used tractors and service, repair, and paint them in your work shop, and then park them out front and sell them."

My father was a smart man, so I decided to give it a try.

On my way home I saw an Allis Chalmers D-12 series II parked out front of a man's house. I bought the tractor and a nice 3 point hitch harrow for $1700.00. I like Allis Chalmers. As a kid that was the first tractor we owned. Dad had bought a "C" before I was born, and I had been breaking land with it from the time I was 9 years old. This tractor (D-12 series) ran like brand new, but it looked terrible, so we painted and serviced it. We sold it and the harrow the next day for $2750.00. I decided to repeat the process, so after getting some more good advice from my Dad, I hit the road.

My father said, "Only buy tractors south of us."

We lived on the Fall Line, so the land to the north of us contained lots of rock, but the land to the south of us was good. This meant that the tractors south of us would be in better condition. As always, he was right.

On my first buying expedition, I bought a 9N Ford, an 8N Ford, three old Fergusons, 2 TO-20(s) and a TO-30. It didn't take me long to understand why I had to pay more for the Fords.

My father had told me to avoid the old Fergusons and instead buy the old Fords. Two of the three Fergusons had bolts that ran all the way through the engine blocks.

After a phone call to my Dad, he remarked, "I told you so."

He explained the old continental blocks used in the 20 horsepower TO-20 and 30 horsepower TO-30 were too light and had a bad habit of cracking down the center of the block between the 4 cylinders.

People back then didn't just throw them away, so they bored holes in the blocks and put long bolts all the way across the block between the cylinders and bolted them up tight.

They then added a can of block sealer and, presto, in many cases the oil and water actually stayed separate.

I studied the tractors as we serviced, washed, sanded and painted them. The only parts that would interchange between the Fords and Fergusons were the tires and batteries, but I had no problem being able to tell that the Fords were much better built.

My father explained that Henry Ford had apparently studied the Ferguson and improved on it, and I agreed.

The TO-20 and TO-30 were still pretty good tractors, small, low profile, wide front end when compared to a mule; this had to have warmed many a farmer's heart. The fact it wouldn't break wind in your face would have been enough to do that.

I was raised on a farm driving a Ford and an Allis Chalmers. My uncle had a John Deere, and we had a neighbor who owned Internationals. I had never had a Ferguson before, but they seemed to be alright. In the late 1970's just before the economy died, I had purchased a Massey Ferguson 230. It was a diesel with power steering, spin-out wheels, with a roll bar and top. It was my only new tractor, but I really liked it.

The Perkins engine ran smooth as silk and had plenty of power. I was big on Fords, but the Massey Ferguson dealer sold me the 230 with diesel and power steering for less than the Ford dealer wanted for a gas 2600 with manual steering and no roll top or spin-out wheels. I compromised my faith in Ford for a nicer machine for the same money. The Massey Ferguson wasn't a sacrifice. After all, I really liked it.

I guess you must be confused about the way I brag about Fords, and yet the only brand new tractor I ever bought was a Massey Ferguson. I learned from the TO-20's and TO-30's that I like the 8N and 9N Fords better, because I felt the Fords were more rugged and better built.

The red and gray colors on the Fords and Fergusons were reversed and many times when I found Fergusons for sale they would be painted the Ford colors. This tells me other people must have come to the same conclusion.

Just after the TO-30's were built, Ferguson and Massey,

a company that built combines, merged to form Massey Ferguson or as many just say, "MF". The Massey Fergusons generally had Continental gasoline engines and Perkins diesel engines. I wasn't much of a fan of the Continental gas engines after seeing so many TO-20 and TO-30 Fergusons with cracked and busted blocks, but the Continentals in the newer TO-35 and MF 35 models seemed to hold up much better.

I guess I was more impressed with Ford, because he built his own engines, and Massey Ferguson bought theirs from somewhere else. Now let me stop right here and warn you about a 4 cylinder diesel called a "standard", or some people refer to it as a freight train engine, because it had a history of being a pill to start. When you did get it running it was practically unstoppable, therefore its name, the freight train engine.

The Perkins engines in the smaller tractors were 3 cylinder, and the standard engine was a 4 cylinder. So, if you are confronted with a TO-35 or MF-35, just count the injectors. Three is good, four is not good. I don't think the average person should tackle the old standard engine.

I would also warn you not to put real money in any of these older MF35 or MF135 or other small Fergusons with something called the multi-power transmission. These tractors have a lever on the right side of the dash, and usually it has a chrome knob on it. This transmission is too old and too complicated for a novice. Heck, I'm no novice, and I wouldn't have one on a platter. These multi-power tractors were really popular when they were new. You could shift the tractor into a faster-slower speed to horsepower ratio without pressing the clutch. The problem with them is when they broke; it pressed the two sides of your wallet together really quickly.

Some people remove the cable trying to fool would-be buyers and put something over the hole in the dash. If you are not sure, be careful. If they were bad then, I would run like a rabbit from them now. Most of these have been crushed or replaced, but still be very careful.

The MF-135 came out in the mid-sixties and except for a

few, still sported the multi- power options. These were some of the finest farm tractors ever built on earth. For instance, the 3 cylinder Perkins engines with the 8 speed transmissions. When accommodated with power steering, this made a heck of a sweet tractor.

Along with the MF-35 and the MF-135, Massey also built the MF-50, MF-65, MF-85, etc. and their newer counterparts the MF-150, MF-165, MF-175, MF-180, etc. The MF-65 and MF-165 and larger tractors had 4 cylinders and Perkins engines. I really appreciate the smooth performance, dependability, and brute strength of a Perkins diesel engine. I still like Perkins engines.

Just recently we formed a small LLC and installed a 4-236 Perkins engine in a full size Ford F-150 pickup. It increased the fuel mileage by 2 to 3 times. It will run over 100 miles per hour and sounds like a Massey Ferguson tractor. If you want to read more about it, go to our conversion site:

www.shadetreeconversions.com

While there, see what rednecks can do when they lose their patience with the greedy oil companies.

Now let's get back to the subject I was talking about. Yeah, I remember. The MF 135 is perfect for the city slicker that moves to the country and realizes that a lawn mower ain't gonna cut it. Sometimes people will run through several lawn mowers before they finally realize they need a tractor.

By the way, if your wife makes all the decisions, take her with you! I would rather be boiled in oil than say something like that, but it's the number one response I get from married men. The women customers just say, "I'll take it. When can you deliver?"

Now I didn't say that I never confer with my little darling wife before purchasing something, but I would never admit it. Come on guys. Act like real men, even if you're not!

Because the MF-135 is now many decades old, you would probably be better off buying something a little newer like the MF230-245-255-265 etc., which were built in the seventies or maybe jump on the MF240-250-270 etc.

For most homesteads of only a few acres, the 40 or so horsepower tractor would fill your bill. If you have cows or horses and want to be able to move the larger round bales, then you need to move up to the 50+ horsepower range (like the MF-165-175 or MF-270-283 series).

You need to be careful with a Massey Ferguson, especially if you mount a loader. The bell housings aren't as strong as a Ford, and the standard agriculture front ends aren't designed for lifting heavy loads. I would suggest that if you need this type of tractor abilities that you consider an industrial version of the Massey Ferguson such as the MF-40B, 40C, 40E or MF-50, 50A, 50C, 50E, or 50F etc. One of these tractors would be a much better choice.

Don't let the yellow paint fool you. Everybody thinks a tractor with yellow paint is an old highway tractor. The truth is the state governments bought only a small number of these. Farmers and small business people bought the lion share of them when they were new.

I need to warn you about something. Some of these industrial tractors have a strange transmission. You step on one pedal for forward and another for reverse. Run like crazy from them. Many have a manual shuttle shift where you still have to press the clutch and pull a lever on the side of the dash to go backwards and forwards. I strongly recommend them over the other system. If the two pedal system goes bad, you will have to refinance your home to pay for it.

Most of the industrial units I mentioned will have factory installed, completely self-contained loaders that bolt to the rear axle on the back and actually brace the tractor and protect the bell housing.

I was in McDonough, Georgia in 1993, and I was waiting to meet a customer to deliver a tractor when I noticed a man unloading a new MF-240 from a trailer behind his pickup. This was one of those lightweight trailers with car tires and an angle iron that runs about two feet high across the front and down the sides. I noticed he was backing off without using any ramps.

Most of these plain little trailers didn't come with ramps; you had to pay extra for them. Many people just made them out of boards. The operator was slowly letting the big rear tires of the tractor to the ground, but after the rear tires landed, he backed up quickly and the front wheels hit the ground so hard the front end of the tractor bounced like a ball. Then I heard a loud POP!

I couldn't believe my eyes. This apparently brand new MF-240 had broken right through the bell housing. I walked over to see what happened. The owner looked absolutely shocked.

"What was that noise?" he shouted.

The man shut the engine down as I pointed to the obvious crack. It had more or less shattered the bell housing. The man got off the tractor and started laughing.

"It's a good thing it's new," he shouted. He then laughed even louder. "The warranty will get a good work out over this," he continued.

I just didn't have the heart to tell him that warranties don't cover breakage.

We had a customer lose a 3000 Ford off the back of a two ton truck running about 65 miles per hour on the interstate. Everybody thought someone else had chained it down. It managed to rock back and forth until it jumped out of gear and rolled off the back. The Ford fell 4 feet at 65 miles per hour and bounced end over end and slid down the median upside down.

We replaced the steering wheel, sheet metal on the hood, radiator, battery, and fenders, but the engine and transmission, including the bell housing, were fine. When I talk about Ford being rugged and sturdy, I'm speaking from experience.

Before we get too far ahead of ourselves, in the early eighties Massey Ferguson imported a small compact tractor from Japan. It was made by Toyosha. If I give you my honest opinion of this tractor, Japan would bomb Pearl Harbor again. If you think about buying one of these 1010, 1020, 1030, 1035, 1040, 1045 tractors, then I suggest you spend a few hours with a couple of bottles of Jack Daniels. By then you wouldn't be able to buy anything. These are one of the sorriest machines to ever be built in Japan. If you bought a Deutz-Allis or White Oliver in this time frame you were in similar trouble.

By the late 1990's Massey Ferguson realized they needed something better, so they made a deal with Iseki. These tractors are excellent, and I would rate them as one of the best imports. Massey Ferguson went from the worst to the best when they made this move in the compact tractor line-up.

Now is a good time for me to explain something. If the tractors you see at the Caterpillar dealers look familiar, then you would be right. The small compact Challenger tractors are Iseki tractors, and the larger tractors are Massey Fergusons. If you walk by the AGCO line of tractors, most of them are the same as a Massey Ferguson.

Anybody who thinks Caterpillar would stop building bull dozers and start putting together small tractors probably believes in the Easter bunny and the tooth fairy. If you see Iseki stamped on the engine of anything, you're looking at a quality product. In today's current line-up of Massey Ferguson tractors, I like the compacts, and I like the 383 and larger models. But, in my opinion, I wouldn't buy a 231S or a 240. I have been inundated with people wanting to sell them with only a few hours on them.

In the late seventies and early eighties, a copy of a Massey Ferguson was built in the Balkans, and it was called IMT. They were a very poor copy at that, but for some reason people will buy anything that is new and shiny. They are like children. If it's pretty and shiny, they think it must be good. Remember the Yugo car? What a piece of junk! After that plant was bombed to pieces in the nineties, an act which I consider a mercy killing, IMT owners were left holding the bag.

If you see something brand new and shiny that has a strange name on it, I would suggest you keep your checkbook in your pocket. If you're going to put your hard earned money into a tractor, new or used, stick with a brand name that you know, and one that has good parts availability. If you do, the tractor will have a good re-sale value. Otherwise, it's like marrying it.

I don't understand why people will buy tractors made in China, the worst of the worst. But if they are new and cheap, they will buy it. Then when it falls apart like a $3 pair of shoes, they bring this junk to us wanting us to be miracle workers; they think we can fix it. We don't waste our time. If we repaired it as good as new, if you could actually get parts, it still won't last but a few more hours. When something is poorly built, it's junk regardless of how pretty it is.

Remember the Yugo and don't be a moron! If you're stupid enough to buy those weird tractors, please call me. I have a great deal on some porcupine eggs.

So to sum up Massey Ferguson, when buying new, you can't beat the compacts that were built by Iseki. I would pass on the 231S and 240 tractors, but the larger ones like the 383 have my respect and admiration.

The only new tractor I ever bought for myself was a MF-230 back in 1976. This was a real quality tractor. I have several personal friends and neighbors who own the MF-135, just remember to avoid the multi-power models although there are some who would fight to their death for them. The MF-165 and MF-175 are also choice machines. I would still buy a new Massey Ferguson from a Massey dealer instead of the Caterpillar dealers. A lot of these deals are those where some brand name or logo is painted on them and sold by someone else. They will fall apart over time, and your re-sale value will go the way of the Goony Bird.

Caterpillar builds some of the best large equipment on earth and leads the world in that area of machinery production, but a MF tractor dealer would be the place to buy your small tractor.

In general Massey Ferguson has played a big part in feeding us and the rest of the world, and I respect the MF label. One footnote when buying a used MF, check the rear seal for leaks and be sure the lift works properly, because they are expensive to repair.

John Deere

I don't understand this global economy. It seems like America always gets the short end of the stick. I don't understand how well-respected American brand name companies like John Deere can go to Europe and buy Zetor tractors, a tractor that I have a poor opinion of. They paint it green and presto, it's a John Deere, a tractor they buy for a few thousand and sell for up to twenty thousand. They also buy Yanmars and paint them green, then sell them as a John Deere model.

John Deere has a reputation for quality that makes them a shining example of American pride. I have been around John Deere from the putt-putt, 2 cylinder M-T to what I believe was not only one of the best John Deere ever built, but probably one of the best tractors on earth, the 4020.

I just don't understand how they could risk their reputation by repainting the Zetors green and selling them to Americans.

Don't get me wrong. I still have faith in John Deere. If I was going to buy a 90+ horsepower tractor, there would be no contest. But personally, I'll pass on the Yanmar and Zetor models. I just hope they know what they are doing.

When I talk to John Deere dealers, they tell me it's almost impossible to build tractors anymore. Between OSHA, EPA, and the unions, they can't survive financially. They import Yanmars for their compact tractors and Zetors for their mid-size tractors. I've been told that most of the larger tractors are built in West Germany.

When it comes to the older John Deere tractors like the old 2 cylinder, tall M-T and 420 type tractors, I wouldn't suggest that an amateur own one of these. They are good quality products, but they are dangerous in the hands of an inexperienced operator. The fact that they are so tall and many have a narrow front end (that means the front wheels almost touch each other) makes them prone to rollovers, especially the ones with hand clutches. If you are an inexperienced operator, leave this model tractor to the professionals.

I think all of our so-called American tractor brands since

about 1970 have been imported. For example, the Japanese brands of small tractors they call "compact diesels" were designed for individuals who wanted something smaller than the tractors being built at that time. I think Kubota started selling so well that they got their attention.

The John Deere 650, 750, 850, 950, etc. are good quality tractors, but they are like all Japanese tractors, which are lightweight. I think almost all the compact diesels are built so lightweight they have to put weights on the front bumper to keep the tractor from rearing up when you let off the clutch. An old Ford, Ferguson or regular John Deere has the steel in the blocks and housings where it should have been, but all in all they are pretty reliable. I do think the "50" series compacts are getting a bit old. Just be careful, the parts prices for these models are like sky high. Be sure that it cranks well. You would have to be sedated if you had to buy an injection pump for one of these tractors. I also tell everybody I don't like loaders or hoes, especially with the shuttle transmission or 4 wheel drive on these tractors. Now the 655, 755, 855 etc, and the 670, 770, 870, 970 are newer, but still be careful before you buy one of these. Drive it awhile and make sure it works well after the engine warms up.

I personally don't recommend the older 1010, 2010, 3010, 4010, etc. tractors, but the 20 series were much better, as well as the "30" series and the "40" series tractors.

I would suggest that you be careful about John Deeres built in France. If you look for an ID plate on the frame, you can usually tell where they are made. Even the John Deere dealers themselves tell me that the French made units don't seem to hold up as well as the West German models.

If you need a tractor with a loader in the 50 horsepower range, take a good look at a John Deere industrial tractor like the 301. It has a heavy front end and is very reliable.

All in all John Deere built some quality products. Their parts prices will definitely get your attention, but fortunately, if you buy a new one or a good used one, they don't generally need a lot of parts or upkeep.

International - Farmall - Case

Back when dinosaurs roamed the earth, the late fifties, and when I was a kid, the Internationals, then known as Farmalls, played a big role in producing food for our nation. Small farmers owned the old "A" models and later "Super A" models. These tractors were small and the driver sat on one side and the engine was on the other. The first ones had a hand lift to pick up the cultivators and bottom plows. This was a time when we had real men. They had to be strong to lift these heavy implements. An exhaust option became available where you pulled a lever and the exhaust from the engine was directed into a tank beside the muffler and the exhaust pushed a plunger inside the tank up. A lever on the outside of the tank picked the plow up off the ground. These old tractors were excellent for cultivating the fields. The driver could follow the plant row as it ran right between his feet.

The exhaust lift was replaced by a hydraulic lift on the Super A and Cub series, a smaller version of the A, about 1955. A few years later they gave it a square grill and called it a "100" and offered a diesel engine.

International also had a "B" series which was an "A" with the engine in the middle. They had a "C", "H", and "M." These tractors were larger and more powerful. The "M" could pull a fully loaded 10 wheel truck out of a plowed field while idling. All of these models became Super series tractors around 1950 by adding a hydraulic lift system. These tractors fed the American people and earned my appreciation and respect.

These tractors were made when second best wasn't an option, but these tractors had big clumsy plows, cultivators, and planter attachments. I don't think today's people would work so hard to use them. These tractors are still in use on old farms all across the country to pull hammer mills and feed mills, generators, and in some rare cases to pull old fashioned belt-drive equipment like saw mills. But, these tractors are not practical for beginners.

Then Farmall built the 300 and 400 series tractor with gasoline engines and had a propane powered option. These tractors were approximately 40-50 horsepower and were built in the mid-fifties. In the late fifties they added a diesel engine option. The 300 became the 350 and the 400 became the 450. They also built an approximately 60 horsepower model called a 650.

In the late fifties and early to mid sixties they introduced the 240, 340, 460 and 504 series from 30-60 horsepower. They were good tractors in their day, but unless you find some older person who calls his tractor "Baby" forget them. The parts prices and up keep would be unbearable.

About that same time, International started selling the B-275 and a little later the, B-414 tractors. These tractors were manufactured in England. They were strong and reliable in the day, but the parts prices again would cause me to avoid them. Anytime parts have to be shipped half way around the world and stored until you need them, the price rises substantially. These models were replaced by the 424, 444, and 544 models. During this period, International built some larger tractors like the 986 and 1086 that were really strong and you will still find them on farms. The farmers that bought them new, in most cases, wouldn't even think of selling them.

Moving up the International food chain, I like the 354 model made during the seventies; but again try to find a low hour, clean, straight one. Remember the parts prices.

In the eighties International re-introduced the Cub series as 254 and 284 tractors. These are light-weight and should only be used for cutting grass. If you put a bottom plow behind one of these, you're just dreaming. These were primarily designed to cut large lawns, and they do an excellent job.

I once saw a loader on a Cub Lo-Boy 154 with a small 4 cylinder gasoline engine and a much smaller clutch. I circled the block to see if my eyes were deceiving me. I couldn't believe anybody would be crazy enough to put a loader on such a light-weight tractor.

As I rode by I saw a "For Sale" sign on it. I just couldn't resist stopping and asking a few questions. I walked up to the door of the home. The tractor was sitting in front of the door. I rang the door bell. This house must have had more dogs than the pound. You could have heard them barking a mile away. I looked back at the Cub 154 Lo-Boy with the loader and thought how could such a puny clutch and pressure plate and light weight transmission survive a loader.

The door opened as a late fifties man wearing shorts fought the dogs off to get outside. This guy had a profound beer gut and 6 inch long chest hair that fuzzed out in all directions. His face was as red as the label on his foot tall beer can. I was glad my lunch had settled. Why do people like that feel a need to show everybody how they look?

When I asked him about the tractor and loader, the first thing he told me was the transmission and/or clutch was out to lunch.

As you could tell I wasn't shocked. I would have been a lot more shocked if he tried to tell me the clutch and transmission were in good shape.

Now don't get me wrong, the Cub Lo-Boy is a great lawn mower and the Hi-Boy model is great for cultivating your garden, but putting a loader on one of these is sheer stupidity! I believe someday I will see a loader mounted on a tricycle or wheel chair!

International throughout the eighties and early nineties did what most other tractor manufacturers did, because our government is hell-bent on stopping us from building anything in the USA. About this time, Case and International merged. Case in point, OSHA and EPA, plus the unions, tempted our American tractor companies into buying something cheap overseas, painting it their colors and then selling it to the public rather than trying to build something here.

If I'm not supposed to say anything bad about these foreign "grab bags" then I'll skip over them, because the same people who bought Ford USA (Fiat) also bought Case-International, and now it's all called Case-New Holland. So, if you didn't like what you saw at the New Holland dealer, except for the red and black paint, then you won't be impressed with what you see at the Case-International dealer either.

I have been told that even the Case backhoe, which I

believe to be one of the top three in the world, will be replaced with the New Holland backhoe. In my opinion the New Holland is the lowest face on the totem pole of equipment. I just hate to see such fine farm equipment being destroyed and replaced with such disdain for quality. Now it's just the almighty dollar. Quality took a hike and corporate profits took over, and as usual we, the people, lose.

A man recently entered our business and proclaimed he had just survived a close call. He explained that he had paid down on a new New Holland tractor only to be warned by a neighbor about how unhappy he was with his and that he was happy to get his down payment back. Then he proceeded to tell me that he went over to the Case-International dealer, and even though he paid $1500.00 more for the same size tractor, he thought he had bought something better.

I didn't have the heart to tell him that he had bought the same tractor painted a different color.

If I sold you a Cadillac and you later discovered that it was actually just a Chevrolet with a Cadillac grill, you could take a warrant for me for theft by conversion or for fraud. You could sue me in court. You would probably want to kick my ---, and I wouldn't blame you for feeling that way.

I don't think it's fair that New Holland sells Fiats and Longs instead of Fords, or that John Deere sells Zetors and Yanmars instead of John Deere, or that Case-International is selling the same old Longs made by Fiat that New Holland sells. I don't think it's honest that Massey Ferguson sells Tafts and Ursus tractors for Massey Fergusons.

According to an attorney who I consulted, he said, "Corporate law allows such practices."

I don't care if it's legal. I don't believe it is a good policy, and that's why I sat down and spent several days writing this book. I feel betrayed by all of this. I don't think it's fair to sell Yugos for Chevrolets or to not be honest with the American people. I would have no problem selling all the different import tractors, if they had both names on the tractor.

I still believe "Honesty is the best policy," but maybe I'm just too old-fashioned.

Kubota

I saw my first Kubota tractor back in the late seventies. It was sitting on the back of a one ton GMC flatbed truck parked at the Lowe's store in Warner Robins, Georgia. I was impressed from the first glance. This little diesel tractor looked like a toy, but it had a water cooled diesel engine and a gear driven transmission. I was used to looking at air cooled gasoline (disposable) powered garden tractor engines with a tinker-toy hydrostat transmission. I wouldn't say it was love at first sight, but I must admit, I really liked the fact that it was diesel, water cooled, and gear driven. This was a farm tractor, but much smaller than anything I'd seen before.

This period in time was during the Jimmy Carter presidency and the economy was in the tank. This was right after my house building business had slowed to a crawl. I was looking for a way to boost my family's anemic income. My father had suggested I start buying small used farm tractors. He said I could service, repair, adjust, paint, and sell them for a decent profit. He was right. It did help and I had a decent work shop with heat, so I could work on them in the winter and sell them in the spring.

I sold probably 35-40 tractors, Ford, John Deere, Allis Chalmers, Massey Ferguson. I enjoyed meeting the people who bought them. One of the best things about the tractor business is that you get to meet "the salt of the earth people." Tractor buyers generally are honest, hard-working property owners.

I had a friend who owned a liquor store, and he always seemed depressed by some of his customers. Selling tractors, I have been fortunate to meet so many fine people during the years I've been in this business.

Unfortunately, it hasn't been as financially rewarding as I would have liked it to be, but meeting so many fine people has been a blessing. Now, let's get back to Kubotas before I have to get a Kleenex.

I went into the tractor business as a full time operation in 1979. I started buying trade-ins from new dealers, just south of me. My father told me to buy them south of the fall-line, because they don't have any rock in the ground to damage the gears. He was right, as usual. When I bought tractors from the northern part of the state, I could tell the difference in wear and tear.

I bought two used Kubota tractors from a dealer in Valdosta, Georgia. This was the first time I ever drove one, but I was impressed. Kubotas are well designed and very dependable. I did have a problem getting my big feet in the tiny running boards, but as time went by, Kubota enlarged them for the American market.

The first ones were small, but today they build them over 100 horsepower. These little tractors were perfect for people with several acres of grass to cut or for pulling a wagon or lightweight grading with a scrape blade. The engines were so fuel efficient that you could drive one all day on a couple of gallons of diesel. They were well designed and they are easy to service. I knew that Kubota tractors were going to do well among the suburban households.

My father told me that after World War II, our government sent Henry Ford to Japan to teach them how to build cars and tractors. He must have done a good job.

As the years went by, I was more impressed with the quality of the Kubota products. For me to brag on a foreign built tractor, I would have to be impressed, because I have never owned a foreign car or truck. I believe in buying American whenever possible.

Our government is run by idiots, and they pushed OSHA, NAFTA, and EPA. At some point in the near future, we won't produce anything with the American label.

Kubota was one of the exceptions I made to the "buy American rule." They have a large plant in North Georgia. That makes me feel better. Kubota now has probably more dealers than any other brand in the southeastern United States.

Before you get the idea that Kubota and I are getting ready to buy curtains together, I will tell you about the few shortcomings they have. Kubota, like all Japanese made tractors, are lightweight. If you buy a 40 horsepower Kubota to pull the implements your 2600 Ford pulled, you'll be in trouble. When purchasing a Kubota to replace your current domestic tractor, buy one that has the same weight. You would need an approximately 65 horsepower Kubota or other Japanese made tractor to pull what a 135 Massey Ferguson would pull. You need horsepower, weight, and proper gearing to pull a heavy load.

My cousin, Tex Watson, is one of the smartest people on earth. He is a house mover, and he will tackle big houses who most people would run from. He explained to me that you have to use heavy equipment like Mack trucks and Caterpillar products to move houses and large buildings. I admire Tex and respect his abilities.

I think Kubota does a good job, but I think these smaller models when equipped with loaders or backhoe attachments are too light for this use. I have already told you in an earlier chapter about the older Japanese man who came by my business and warned me not to put loaders or backhoe attachments on these smaller tractors. He told me they weren't designed to use them. In Japan they use something like a skid-steer on tracks for loading and excavators for digging.

I explained to the man that I didn't put the attachments on that particular tractor. I was only selling it on consignment and that I agreed with him.

He just bowed his head, got back into his Chevrolet S-10 and drove away.

When you see these little import tractors with something heavy on the back, they generally have a bunch of weights on the front to add additional weight. If they had made the housings, blocks, crank shafts, and gears as heavy as a Ford, you wouldn't have to put weights on the front. It's kind of strange. They made them so lightweight that you have to pay for weights to replace the metal they left out during the design and building process.

I would never put a loader on any compact tractor under approximately 40 horsepower. The front end is too light. You need power steering, if you want to be able to steer it when the bucket is full. Four wheel drive front ends have a heavier front axle, but I have seen them break where they mount under the front end.

I know the next thing you're going to tell me is you don't plan to put anything heavy in the bucket. Sooner or later you or your neighbor, Bubba, will put something heavy in it.

If you buy a used compact tractor with a loader, you can use the loader to pick the front end off of the ground. Then you can grab the front wheels and jerk them back and forth to check for wear and tear in the front end, steering, linkage, and wheel bearings. If it's loose as a goose, it will have to be fixed. Take that into consideration during your negotiation of the purchase price.

I would suggest if you are determined to buy a used compact tractor and want a loader, you should buy a good used tractor with power steering and positive traction rear end, and then have a good used or new loader installed. The $500-$1000 you pay will offset the additional wear and tear to the front end, steering, clutch, brakes, and drive train.

Some of the first Kubotas that I drove were the old two cylinder models. I don't like the shake and vibration of the two cylinders. I suggest you buy a three cylinder or larger engine model. The three cylinder models are much smoother. When buying a new or used Kubota, I am not a 4 wheel drive fan. Some people believe a 4 wheel drive 25 horsepower tractor will pull as well as a 50 horsepower 2 wheel drive. Only in your dreams, remember the issue of weight.

If you buy a used 4 wheel drive, be sure to check for loose play in the front end. If you buy a new 4 wheel drive tractor, be sure to only engage the 4 wheel drive on rare occasions. Don't operate in 4 wheel drive unless you need it. If the 4 wheel drive goes bad on your new tractor, it will probably be after the warranty runs out. The repair bill on these front drive units can get your attention. That's why I'm trying to warn you now!

I also wouldn't suggest the shuttle transmission, especially with a loader. The manual transmission is much stronger and will hold up much better. If you're such a sissy that you can't drive a manual transmission, maybe you will become more of a man when you have to pay several thousands of dollars to repair the shuttle.

Always stop your shuttle transmission with the brake pedal before changing directions. If you don't, you will pay the dealer's repair shop for your laziness.

I suggest that you buy a Kubota that is large enough to properly perform the work you need to do. Too many times people call me wanting to trade up to something bigger and I ask them why they bought something too small. About 98% of them say that it was because of the price.

If you need a 40-50 horsepower tractor, a 20-25 horsepower unit won't do the job. It is much better to have a larger payment now than to have to trade up later on. If you plan to operate a 5' rotary mower, you need at least 35-40 horsepower. If you plan to use a 6' rotary mower and pull bottom plows, harrows, and operate a posthole digger, you need approximately 50 horsepower when buying a Kubota. When you try to get by with a smaller one, it will be too much stress and strain on the running gear.

I suggest that you refrain from loaning your tractor to other people. Over 90% of the physical damage we repair in our shop occurred after someone loaned their tractor to a neighbor or friend. It would be much better for you to mow their field for them rather than loan it out. Most people who borrow a tractor couldn't care less about the damage they do to it.

Service is important to any tractor, but I believe it is especially important to a compact tractor, because people tend to overwork them. A larger tractor can better handle the hard work.

If I was going to buy a brand "new" tractor in today's market, I would buy a Kubota. I would buy a 2 wheel drive, manual transmission model, large enough to do the job. Kubota has an outstanding reliability record and a good resale value to go with it. It's always important to buy something that will hold its value. You never know what the future holds.

While I was writing this book, a customer of ours called to

ask us to pick up his Ford 4600 and equipment and re-sale it. Apparently he had lost his wife after retiring and coming home. Now the only family he has is his children and grandchildren. They live almost 200 miles away. We should be able to re-sale his tractor and equipment after servicing it and making a few repairs for approximately 90-95% of what he paid for it. That's why I only sell quality brand name equipment. If he had bought the farm tractor he was considering when he bought the 4600 Ford from us, considering the fact that Farmtac USA just went bankrupt, he could have lost 90% of the purchase price.

 I prefer quality name brand equipment like John Deere, Ford, Massey Ferguson, and Kubota. To the people who say they cost too much, my response is "You get what you pay for." The market today is flooded with a lot of what is called "gray market tractors" including Kubotas. These tractors were sold new in Japan and generally have as much as 10 times the hours or wear and tear. They are bought used in Japan and shipped here. Most importers of these units paint them as pretty as a circus pony. Some people are actually fooled into thinking they are new. One way to tell a domestic Kubota from a "gray market" one is to look at the PTO lever. Most of the foreign made ones have several PTO speeds. Also double check the decals on the dash and engine for Japanese writing. The Kubota dealers generally bad mouth these units, but high hours is what I worry about.

 In Japan these tractors are worked around the clock till they drop. Then they are shipped here for re-sale.

 I need to warn you about fake Kubotas. I had to look twice at some tractors on a flatbed semi truck one day. They were painted just like Kubotas, but they were actually Yanmars. I couldn't believe it. Some crook had gone to the trouble to have decals made that said Kubota, but they fit the shape of a Yanmar hood. It never ceases to amaze me, the efforts a crook will go to in order to cheat somebody.

 I talked to the truck driver on my CB radio. He told me he was delivering them to a used equipment dealer, just north of Atlanta.

I asked the driver, "Do you know what kind of tractors you are hauling?"

He just laughed and said, "I don't care. I just drive the truck."

Do you remember at the start of this book when I told you how corrupt this business is?

If you are not sure about a tractor, copy the serial numbers on the identification plate, and then call your local dealer. If it doesn't have a plate, then call your local sheriff. To sum up Kubota, I would say, buy the right tractor for your job. Service it. Don't loan it out and take good common sense care of it. A shelter would also protect your investment.

On my scale of quality for new tractors, Kubota is the only one that rates a "10".

Allis Chalmers

My first tractor ride was with my Dad on his Allis Chalmers "C". The "C" had a narrow front end, and they also made a "B" with a wide front end. The "C" was a two row tractor and the "B" was a one row. They also made a "G" model. It was a one row tractor with the engine in the rear. People fight over these.

I still remember sitting in my father's lap and steering that old "C". These tractors were built from the mid-forties to the mid-fifties. These tractors were solid, strong, and dependable. The engines were practically indestructible. I have many fine memories driving that old Allis back when I was only a kid.

We pulled a 4 wheel farm trailer through the woods usually at night to pick up tar barrels and transported them back to a loading dock where we loaded them on our ton and a half Studebaker truck. I loved that old tractor, but it didn't have a lift, so Dad moved up to an 8N Ford in 1957.

They were replaced by the "CA" and by larger models "WD" and "WD45".

I still know people who own these old tractors, and they won't sell them for love or money. These were really great quality tractors. They were replaced by more modern looking tractors in the sixties, the D-10-12-14-15 and 17's. These tractors had the same strong engines, but many of them developed transmission problems over time. It was imperative to keep the lube clean in the transmission. Later a D19 and D21 were added to this line up as Allis Chalmers built larger tractors to keep up with the competition.

In the late sixties and early seventies they came out with the 170-180-190 and the 190XT tractors. These carried on the legacy from the forties of good quality tractors, but John Deere came out with its 4020 version in the mid-sixties. The 4020 John Deere caught everybody in the tractor business sleeping.

The worst part of the old Allis Chalmers tractors was the transmissions. You had to be careful shifting them and keep them full of lube.

In the late seventies to the mid-eighties, Allis Chalmers

Allis Chalmers

followed the other tractor companies and imported a Japanese compact diesel line. Some held up. Some disintegrated. Be careful.

The 5020 and 5030 were next and easy to operate, but if you needed a part, you would feel like you gave blood and didn't get your milk and cookies.

The 5040 and up is what got my goat. Allis Chalmers just bought Longs and painted them orange and changed the grill. My worst nightmare is that I'll die by being run over by a run away Long tractor while sitting on a park bench in front of the old folk's home.

I hate Longs. I don't give a ---- what color you paint them. In my opinion they are like a disease that can't be cured.

The New Hollands are Longs. Case-Internationals are Longs. These older Allis Chalmers are Longs made by Fiat. Obama just gave Chrysler to Fiat. I would have rather Chrysler died peacefully than to be prolonged in misery.

The Chinese made tractors and the junk made in India are the only tractors worse than a Long.

A good friend of mine and fellow equipment dealer calls Longs "misery on wheels."

I agree. The only way to fix a Long is to melt it down to a big blob and use it for a shrimp boat anchor.

Yanmar

These small diesel tractors known as compacts were sold in the USA back in the seventies. I know several people who bought them and still have them. Like all the small diesels out of Japan, they were lightweight, but seemed to hold up well as grass cutters.

The worst problem I saw with them was parts supply. The wholesale distributor in our area didn't have a parts supply other than filters and hoses. I know one man who waited over a year for a clutch.

Just as suddenly as they appeared they seemed to disappear in the late seventies. I was told by a friend of mine who sold John Deere that John Deere had made a contract with Yanmar to build the small compact John Deere. I guess that is why the Yanmars weren't being sold anymore. In our area I didn't see another Yanmar for years.

About the mid-nineties they started showing up at tractor auctions by the hundreds. I thought they were new until I examined them more closely. This was the best job of making a used tractor look new that I had ever seen. They even painted the screws, belts, hoses, knobs, and tires. I realized that someone had gone to a lot of work to fool people into thinking they were new. The 3" deep cleats on the tires really stood out.

I tried to start one, and the little two cylinder diesel knocked like a train for the first minute or so. This poor tractor had more lose play in the steering, pedals and linkage than I had ever seen before. I wondered why somebody would go to so much trouble to make these tractors so pretty and yet not try to fix the mechanical problems.

The individuals at the sale were completely fooled and salivated all over themselves bidding them up. I couldn't believe how these people were fooled by the shiny paint job. Even though some were knocking and some were smoking, the bidding frenzy continued. The dealers weren't so easily fooled and just watched in amazement.

The buyers had big smiles on their faces after buying these worn out tractors.

I tried to warn some of them, but they wouldn't listen.

One month later I returned to the same auction sale, and the scene was completely different. Several of the painted Yanmars were on the salvage row, and several others were scattered throughout the tractor rows.

I realized that one of the men who was selling these tractors was a friend of mine, so I asked him about the tractors.

He told me that they were buying them in Japan and shipping them to the USA in containers. He explained in Japan farmland is very scarce and that as many as five families will go together and buy a couple of acres and pay as much as a million dollars or more for it. Those families will share one tractor. Generally Yanmars are the lowest priced tractors in Japan, so they run one small tractor 24/7 for about 20 years. Remember what I said before. Yanmar makes good equipment, but some of them will have 20,000 hours or more. These people know that after 20 years it's hard to get parts for these little tractors, and they know they're not worth rebuilding, so they buy a new one and sell the old one to an international wholesaler.

So, my hypothesis was right. Yanmar is a good product, but with the enormous amount of hours these machines have, they are generally worn out. So instead of rebuilding these worn out tractors, they simply make them beautiful instead. The average person is so excited about the meticulous paint job that they don't take the time to check them out mechanically.

We paint our used tractors, but we fix them first. Many people in the used tractor business say it's cheaper to make them pretty than to fix everything.

I was at another used dealer's lot above Atlanta one day and noticed a Massey Ferguson 135 with a puddle of oil under it on the pavement. The rear seal was so bad that the engine oil dripped out around it. When I pointed it out to the owner, he just moved the tractor over to a portion of the lot that wasn't paved!

We would have replaced the seal even if we had to charge more for the tractor. We had to acknowledge that he must be right, because he drives a Mercedes and I drive a Ford.

On a recent trip to a wholesaler's lot in Alabama, the owner was trying to get me to buy a used Yanmar tractor. I had purchased a dozer from him and was waiting for them to adjust the steering clutches. He insisted his used Yanmars were better than the other "gray market" imports. I examined 30 and found major mechanical problems with all of them.

Another problem you have with the used Yanmars is parts. Everybody who sells them will tell you that somebody else has the parts. You better check on the parts for any strange tractor you're thinking about buying.

Our market today is flooded with all sorts of strange import tractors, and I don't suggest you buy them unless you're sure parts and service will be readily available. In most cases, a few years down the road you will be left holding the bag.

A tractor wholesaler stopped by my business one day. He had a red tractor on his trailer behind his pickup. It was called a Branson. I admit I was impressed with the way it looked. The salesman told me it was a copy of a Kubota made in Korea.

I asked him, "What is the wholesale and retail price of the tractor?"

He looked in a notebook and quoted me the prices.

I looked at the figures and asked, "Why would anybody pay the same money for a tractor that nobody has even heard of instead of buying a Kubota?" I added, "Kubota has an outstanding reputation and excellent re-sale value. It also has a nationwide parts and service network."

He closed his book and said, "Well, if that's the kind of attitude you have, I know I'm wasting my time here!"

I don't see what attitude had to do with it. It was common sense, but he drove away and never looked back.

With three Kubota dealers within twenty miles of me, I wouldn't think of tangling with a Branson unless that newcomer had something special to offer like a price discount. I couldn't count the number of people who have bought a small tractor, only, from me. When I ask about implements,

Yanmar

I know what they're going to say.

Their response is, "I already have the implements. I bought one of those Yanmar package deals."

The latest gimmick for selling not only the "gray market" Yanmars, but Mitsubishis, Isekis, and Hinomotos is to form a package deal. You see them advertised sitting on a trailer with a mower and scrape blade.

I asked one of the dealers who sell them that way, "How does that work?"

He said, "We pay extra for a couple of these tractors that we use as demonstrators around the lot."

"They are in much better condition than the ones on the trailer, the package deals," he chuckled.

I got the picture. If you buy a package deal, make them unload the tractor sitting on the trailer and be darn sure you drive it before you sign the papers. Also be sure to read the fine print on your contract. Some have a 95% pre-payment penalty. This means even if you could re-sale it to get back part of your money, you will have to pay almost all of the interest anyway.

Personally I wouldn't buy one of these package deals even if I was drunk!

I always recommend that you go to your local credit union or bank and pre-arrange financing before you buy. It is never a good idea to finance with some outfit you have never heard of. Generally speaking, your local credit union or bank won't have all the "gotchas!"

Another thing guys, never admit you're such a sissy that you have to go ask your wife before buying something. Try to have some pride. Just say you want to think about it, or better yet, bring her with you.

In conclusion I'm going to sum up Yanmar by saying they build good products, just don't buy a worn out one, especially in the package deal from hell with an on the spot finance deal the devil would love.

Oh, by the way, many of these painted up, gray market Yanmars, will knock after the first oil change. That's because they had 90 weight gear oil in the crank cases. In this situation, the extra thick oil will help prevent the start up knock. If you check the fuel tank, you will usually find some gasoline mixed with the diesel, so the worn out engine with the low compression will at least start.

Special note: Look at the International Cub Cadet diesel. Have you noticed that they look a lot like the Yanmar made compact diesel tractors? This global economy is a strange thing.

I have faith in Yanmar products, but just be sure when you buy one, thoroughly check out the condition. Don't put all your faith in some sort of warranty. A warranty is only as good as the person you buy from. Some warranties have more holes than a sponge.

I have had grown men stand in front of my sales counter and cry like children about getting the shaft. Many of these people get into a mess by trying to be tight with their money. I have had thousands of people stop by my lot driving a $40,000-$50,000 car or truck wearing $5000 worth of clothes and jewelry and want to buy a good used tractor for $2500 or less.

It takes a certain amount of money to buy something that will last you the rest of your life. If you buy a clunker trying to save money, in a few years you will have more money in repairs than if you had just got off the wallet and paid a little more.

Long-Fiat

I have already explained the connection between Long-Fiat and New Holland. I don't like a Long tractor, even if it is a short one. They sold some of them in this country in the sixties and seventies. People bought them just because they were cheap. Just remember, you get what you pay for! This is just our opinion, of course, but many people must agree with us, because of the dismal resale value.

Longs are awkward looking and difficult to drive. More importantly, almost nobody will work on them, and the parts are so expensive. When they have to be repaired, many of them become shrimp boat anchors! Unfortunately, they are being sold today as New Holland and also as Case I-H tractors. I couldn't count the number of individuals, county commissioners, and city councilmen who have called me and complained about Long, New Holland, and Case I-H tractors.

In my opinion they don't perform to the standard that most people expect from a new tractor. Now I'm sure if you ask one of their dealers, you will get a somewhat different opinion. I suggest that you talk to several people who already own one before you listen to me, a compulsive bull-shipper, and make a decision for yourself.

White Oliver

Oliver built some really good tractors up until the seventies, but then they started painting the Fiat-Long tractor green and white and turned my stomach. They imported a Japanese compact tractor that I have very little respect for. I remember the 550 about the size of a 135. It had a live PTO shaft and was a really good tractor. It was easy to handle, but I lost interest when they became Fiats and Toyoshas.

It's a hot day and I'm sitting at a picnic table in front of my shop under a big shade tree. Next door a man is cutting the grass around a small country church with a zero turn radius mower. He has already had two mechanical problems and now he's asking for help again.

He said the machine has just run out of warranty and he's had nothing but trouble.

Do you remember what I said about the zero-turn-radius mowers? Why am I always right? Well, when it comes to mowers, that is.

Belarus

I don't know why I'm using up perfectly good ink to write about the Belarus. It's a Russian made tractor. They sell cheap and people, as I have said, will buy anything cheap and shiny.

A man came to my shop late one winter afternoon and told me his father gave him his Belarus tractor.

I felt sorry for him. He must not have cared much for him. Buying a Belarus is like marrying a really ugly woman. You are stuck with it for life. I have to live down the shame, because I bought some of these tractors new.

My older brother wanted one to use on his feed lot. The Russians were apparently the first to use 4-wheel drive, and he wanted something that he could use to feed his cows. In rainy weather this could be a real challenge.

I ordered four of them. I couldn't believe how rough and outdated they were. The engines were the most redeeming part of the tractor. The first ones had centrifugal oil filter systems. I actually liked this feature. When the engine turned 2000 RPM's the oil filter drum spun at about ten times the engine speed and the trash and dirt was pressed hard against the walls of the spinning drum. Then you just took it apart and cleaned it out. This kept the engine oil cleaner than any filter that I had ever seen.

So, Belarus, even though it was crudely built, had a reasonably tough 4 wheel drive, pretty good engines, and an oil filtering system that I liked. However, in my opinion, if you buy one of these, you need your brain examined.

They have a terrible re-sale value. Some of the earlier machines warned you not to pull anything heavy in low range. So, if you're going to pull something big and dangerous, be sure to drive in high range. It explained that pulling something heavy in low range could damage the transmission. It stands to reason that if you buy one of these tractors, you have moved from the group of people who need to be watched and now you have joined the people who need to be looked after.

Gray Market Tractors

These are tractors sold new primarily in Japan. They are remarkably similar to the ones sold here, except for tiny running boards, Japanese writing on the dash, and various warning plates and several speeds on the PTO. In Japan these tractors are generally worked to death. They may have 10,000 to 20,000 hours or more.

A similar new tractor sold here in the USA when it is 20 years old may have only a few hundred hours. Here they are used to cut grass for people who have a large lot or a couple of acres.

In Japan up to five families may buy one together and run it around the clock until it's on its last legs. Then they are crated up and shipped to the USA. They are painted so pretty that some people think they are new, but generally speaking they are worn out.

The dealers are jealous that they didn't get to sell them new, and so they talk about them like they were red-headed step children. Most, if not all of the parts, will interchange with models sold here. For example, Kubotas can be a gray market tractor, but use parts interchangeable with normal Kubotas.

But, if you buy a Yanmar, Mitsibishi, Hinamoto, Iseki, or Shibaura, your chance of finding parts and a leprechaun at the end of a rainbow are about equal. Basically, these tractors are worn out and need to be buried. If they were worth rebuilding the owners would have done so. Instead they are shipped halfway around the world to some unsuspecting pigeon hunting something cheap!

Don't be a pigeon or a fool! Your wife, your boss, and the government will bleed, humiliate, and enslave you enough. You don't need a tractor-from-hell deal to boot!

Many of these tractors are sold sitting on a trailer with a mower and a scrape blade. It's called a package deal. If you buy one, add up what the implements and trailer are worth. This is what you are buying. The tractor though painted and pretty will be the least valuable part of the deal.

The dealer will preach that they have a warranty. Remember what I told you about warranties? They are only

Gray Market Tractors

as good as the person you buy from. Any time you are in doubt, call the Better Business Bureau or State Consumer Service Department where you live.

I have read some of these contracts, and the warranties that I read are useless. I can't count the number of people who bought some of this junk who have called me for help. I can't help you after you have made a mistake of this magnitude. Some of the contracts I read even had a 95% pre-payment penalty. This means if you realize how bad a deal you got and try to sell and pay off this package deal, you will still have to pay off years of interest that usually runs over 28%. I have seen some that were as high as 48%.

I guess some dealers got the idea to offer package deals. If the deal is on a new tractor with a quality brand name, you're probably OK. I suggest that you finance through your local bank or credit union.

I can't get over the number of people who are too lazy to go to the bank. They finance at "on the spot" lenders, or equally as bad, they put their purchases on credit cards. Let me insert here. If you use a credit card, call the card company before you buy. You can negotiate the interest rate in advance. I have had a customer get an 8% loan from a credit card by calling in advance and getting the purchase pre-arranged.

There are several companies across the country that sell "gray market" parts, but the prices will get your attention. I know of a case where a man ordered a head gasket for a Mitsubishi tractor. It cost him $150.00. A head gasket for a Ford 2000 is only about $15.00. When his gasket came in, it had gotten bent in shipment. For those of you who don't know, a bent head gasket is as worthless as a used toothpick.

Now to be fair, I'm going to tell you about my experiences with Komatsu "gray market" machines. I have sold several dozers, tract loaders, and excavators made by Komatsu. Komatsu is a good product, and I respect their quality. The auctions on some occasions are full of these and other Japanese "gray market" used machines sold new in Japan. I have owned many of them, and they are every bit as good as the machinery sold here.

When the value of our dollar is soaring, you will see equipment speculators bringing these machines into the USA. When the American economy gets a cold, the saying goes, Japan gets pneumonia. Most countries hate to admit the fact that the USA is the financial center of the world. Even some of our enemies admit this, but few of our allies do. So when the dollar is high, you can buy "gray market" machines cheap.

My conclusion is that normally dozers and excavators won't be as worn out as the little tractors, but you need to be careful when buying them. You should determine their condition on a one by one basis.

Conclusion

We wrote this book to help people that didn't have the good fortune to be raised on a beautiful farm in Dodge County, Georgia, like us. With our combined 100 + years of experience with farming and farm equipment, we have been blessed to enjoy the beautiful outdoors. If we had been office workers, I don't think we would have survived. Being born on a farm and being a building contractor for many years in addition to being in the equipment business for over thirty years, I feel blessed that God gave us the pleasure of being able to enjoy the great outdoors.

Our constant involvement with tractors and equipment has been a wonderful blessing. We have enjoyed it. It has been a lot of hard work. The rewards of a job well done, and being in contact with the most interesting, salt of the earth people, has made this life well worth it.

Whether it was producing timber, cattle, or turpentine on our farm in Dodge County or building homes for families in Warner Robins or selling tractors, our goal has always been the same, to do the best job we could to provide the needs of others.

We always prided ourselves in putting the customer first. I think today too many businesses have forgotten that principle our father taught us so long ago. The tractor business has gotten so competitive that even we have had a hard time keeping up with it, but the part that upset us was when we saw people being taken advantage of by unscrupulous people. When we heard some factory representative of an equipment company bragging about how to take advantage of every day people's ignorance of the equipment world, we felt compelled to write this book.

We are not saying everybody in this business is corrupt. Far from it, but the newest wave of corporate drive seems to be more and more corrupt. We still believe in the American people and the God that blessed us to live in this great country. We believe all people need is a fair deal and honest dealings and they will succeed, but to use your knowledge to take advantage of others is wrong.

When somebody walks in our door and says, "I don't know

a thing about equipment." We are grateful they came to us. No matter how much work we need to do, we feel it's our responsibility to do our best to give them a reader's digest version of this book. Even if they buy from somebody else afterwards, we feel obligated to share as much information with them as we can to protect them from the few that would take advantage of them.

Whether this book sells a few or thousands, we felt obligated to write it. If you know someone that might be looking for a tractor, please tell them about this book. You may be their safety net. We know our country is in a heck of a mess, and people don't need to waste a dime. Right now you don't have to own a tractor. But, if things continue to get worse instead of better, owning a tractor could mean the difference between feeding your family or not. Then this book would be more important than ever.

Regardless of what the future has in store for us, I think the Almighty wants us to work together and help each other. If we do this, we will survive and raise our families just like our forefathers would expect of us. We can carry the torch with strong hands and good hearts and pass it on to the next generation.

It worries me that too much emphasis has been placed on the paid off gold diggers running Washington. We have been blessed to meet and become friends with thousands of the best people on earth, our customers.

The most important rule for buying a tractor follows:

Simple is best! The best tractors are rugged, simple, heavy duty and reliable.

Many of the new tractors have computers. Almost all of them are very complicated. I asked a New Holland dealer shop foreman if he could fix the CVT (constantly Variable Transmission) in the new 8N Ford Heritage look-a-like.

He just shook his head and said, "I don't think anybody could."

Conclusion

These new tractors with all these super duper automatic transmissions are "junk yard bait."

Dealers tell me that first time tractor buyers are too lazy to step on the clutch between forward and reverse. If you are that lazy, you don't need a tractor. If you buy an automatic or shuttle shift tractor, you will have to mash the brakes between forward and reverse or you will become a regular patron of your dealer's garage.

My advice is, avoid complicated tractors. The best tractors have the fewest moving parts. The more parts the more wear, the more chance of mechanical failure. The greater chance you will be financially plucked.

If you look at a tractor under 60 horsepower and it has steel weights on the front, this is a clear sign of an engineered failure. If the builders of these tractors had built the engine blocks, crank shafts, rods, pistons, bell housings, line shafts, gears, axles, rear housings, and everything else heavy enough, they wouldn't have to put weights on the front. So when you see weights on the front, remember you're paying them for steel that the manufacturer short-changed you of when they built the tractor.

Look at the Ford 2000-3910's or the older John Deeres, Internationals, and Massey Fergusons. All of these tractors were simple, rugged, heavy duty, reliable, strong, and durable.

As consumers we should demand heavy duty, quality-built products. When you buy junk you encourage the production of more junk!

Wake up! Don't be a sucker!

Notes

Notes

Notes